T0139005

Intelligent Systems for Stability Assessment and Control of Smart Power Grids

Yan Xu

Nanyang Technological University
Singapore

Yuchen Zhang

University of New South Wales
Sydney, New South Wales, Australia

Zhao Yang Dong

University of New South Wales
Sydney, New South Wales, Australia

Rui Zhang

University of New South Wales
Sydney, New South Wales, Australia

CRC Press
Taylor & Francis Group
Boca Raton   London   New York

CRC Press is an imprint of the
Taylor & Francis Group, an **informa** business

A SCIENCE PUBLISHERS BOOK

CRC Press
Taylor & Francis Group
6000 Broken Sound Parkway NW, Suite 300
Boca Raton, FL 33487-2742

© 2021 by Taylor & Francis Group, LLC
CRC Press is an imprint of Taylor & Francis Group, an Informa business

No claim to original U.S. Government works

Printed on acid-free paper
Version Date: 20200718

International Standard Book Number-13: 978-1-138-06348-8 (Hardback)

Visit the Taylor & Francis Web site at
http://www.taylorandfrancis.com

and the CRC Press Web site at
http://www.routledge.com

Foreword

Stability is an essential requirement for the power system. Since the new century, modern power systems have seen large-scale integration of renewable energy resources such as wind and solar power, and more non-conventional loads such as electric vehicle charging loads and those support demand response programs. These new elements, due to their different dynamic characteristics and/or high-level randomness, make the power system stability assessment and control a much more challenging task. The conventional model and simulation-based methods become inadequate to adapt to this new environment. In the meantime, with the wide-spread deployment of phasor measurement units (PMUs) and other metering technologies, more measurement data of the power system become available, which opens the opportunity for data-driven stability assessment and control.

This book presents a series of intelligent systems that make use of measurement data for stability assessment and control. Compared with conventional methods, the intelligent system solution has a much faster assessment speed, less reliance on system models, and a stronger ability to provide decision-support knowledge.

Written by a dedicated research team who has been working on this field for over 10 years, this book offers systematic coverage of state-of-the-art intelligent systems for data-driven power system stability assessment and control. It begins with the introduction of power system stability, offering readers a broad and stimulating overview of the field. It then identifies the problems and difficulties of conventional methods for stability assessment and control in the new environment and justifies the need for the intelligent system solution. Subsequently, the book introduces the framework and major blocks of an intelligent system for stability assessment and control, walking readers from introduction and background to general principle and structure of the intelligent system. After this, the book digs down into technical details on on-line stability assessment and preventive stability control which focus on the pre-fault state of the power system, and real-time stability prediction and emergency control which focuses on the post-fault state of the power system. The authors demonstrate a variety of innovative methodologies for stability database generation, feature extraction and selection, machine learning model design and training, decision-making and result credibility evaluation, parameter optimization, etc. To illustrate the concepts and methodologies, simulation tests on different benchmark testing systems are also given. Finally, the book moves to missing-data issues that often appear in practice and would impair the intelligent systems. Effective methods to address such issues are introduced.

Intelligent Systems for Stability Assessment and Control of Smart Power Grids is a systematic presentation of the authors' research works and their insights into this field. With a balanced presentation of theory and practice, this book is a valuable reference for researchers, engineers, and graduate students in the areas of power system stability assessment and control.

Dr Innocent Kamwa
Fellow of the IEEE
Fellow of the Canadian Academy of Engineering
Chief Scientist for Smart Grid & Head of Power System and Mathematics
Research Institute of Hydro-Quebec (IREQ)
Varennes, QC, Canada

Preface

Modern power systems are evolving towards the Smart Grid paradigm, featured by large-scale integration of renewable energy resources, e.g. wind and solar power, more participation of demand side, and interaction with electric vehicles. While these emerging elements are inherently stochastic, they are creating a challenge to the system's stability and its control. In this context, conventional stability analysis tools are becoming less effective, and call for the need for alternative tools that are able to deal with the high uncertainty and variability in the smart grid.

In the meantime, Smart Grid initiatives have also facilitated wide-spread deployment of advanced sensing and communication infrastructure, e.g. phasor measurement units at the grid level and smart meters at the household level, which collect a tremendous amount of data in various time and space scales. How to fully utilize the data and extract useful knowledge from them, is of great importance and value to support the advanced stability assessment and control of the smart grid.

The intelligent system strategy has been identified as an effective approach to meet the above needs. After over 10 years' continuous research in this field, we would like to present our intelligent system solutions to power system stability assessment and control in this book:

In Chapter 1, the preliminaries of power system stability are reviewed, including its definition, classification, phenomena, and different stability indices.

In Chapter 2, the problems and difficulties of stability assessment and control are discussed, and the motivation for the intelligent system solution is described. We classify the problems into two types: on-line stability assessment and preventive control, and real-time stability prediction and emergency control.

Chapter 3 introduces the general framework and the major steps for developing a practical intelligent system for stability assessment and control. We also give our insights into the key challenges and discuss our philosophy for addressing these challenges. The developed methodologies in the following chapters are based on this philosophy.

Chapter 4 presents the intelligent system for on-line stability assessment, which aims to use steady-state operating variables to achieve fast stability assessment for potential contingencies. We had originally proposed the credibility-oriented methodology to evaluate the reliability of the output of the intelligent system. If the output is reliable, it can be directly utilized for stability assessment, otherwise alternative methods such as traditional time-domain simulation can be used to replace this result. In such a way, the practicability of the intelligent system can be significantly enhanced. The detailed methods and algorithms for stability database generation, feature selection, machine learning model training, credibility evaluation criteria, ensemble-based decision making, and simulation test results are provided.

Chapter 5 presents the intelligent system for preventive stability control. We aim at transparent and interpretable preventive control actions which manipulate system operating state to counteract possible contingencies. The key methods and algorithms for stability knowledge discovery, interpretable control rules, as well as simulation test results are detailed.

Chapter 6 presents the intelligent system for real-time stability prediction, which aims to use real-time synchronized measurements to foresee the stability status under an ongoing disturbance. In order to balance the real-time stability prediction speed and accuracy, we had originally proposed the time-adaptive decision-making mechanism which can progressively predict the system stability based on available measurements. The detailed approaches for machine learning model design, time-adaptive decision-making process, optimal model parameter tuning, as well as simulation test results are provided.

Chapter 7 presents the intelligent system for emergency stability control. We aim at fast decision-making on stability control actions at the emergency stage where instability is propagating. The key methods and algorithms for emergency control database generation, model input, and output selection, as well as simulation test results, are detailed.

Chapter 8 aims to address the missing-data issue that usually appears in practice and is a critical challenge to the intelligent system. We had originally developed novel methods based on power grid observability and the deep learning technique. Detailed methods for feature selection and clustering, machine learning model training, and decision-making, as well as simulation test results, are presented.

We hope this book serve as a timely reference and guide for researchers, students, and engineers who seek to study and design intelligent systems to resolve stability assessment and control problems in the smart power grids.

We would like to sincerely appreciate the funding agencies and institutes who have supported our research in this area, listed in alphabetic order: Australia Research Council (ARC), Electric Power Research Institute (EPRI) in USA, Hong Kong Research Grant Council (RGC), Nanyang Assistant Professorship (NAP) from Nanyang Technological University, National Nature Science Foundation of China (NSFC), Singapore Ministry of Education (MOE), University of Sydney Postdoctoral Fellowship.

Yan Xu
Yuchen Zhang
Zhao Yang Dong
Rui Zhang

Contents

Power System Stability: Definitions, Phenomenon and Classification

Power system stability has been recognized as one of the most important problems for secure system operation since the 1920s. Many major blackouts caused by power system instability have illustrated the importance of this phenomenon. This chapter introduces the basic definition and phenomenon of power system stability.

1.1 Overview

Over the last two decades with the steadily growing population, the consumption of electricity is also continuously increasing all over the world. Specifically, the global electricity consumption has increased by three per cent annually over the last 20 years (Y. Enerdata, 2017). Moreover, the concerns of global warming and its induced climate changes have pushed the increasing use of renewable energy sources (RES) in many countries to improve the energy sustainability. Many government authorities have announced their long-term targets of increasing the RES penetration for future power grid to overcome the current energy challenges. For example, the Australian Clean Energy Regulator has announced a 20 per cent renewable generation portfolio target of Australian electricity supply structure by 2020 (C.E. Regulator, 2012); the U.S. government has released the target of achieving 50 per cent of renewable power generation by 2025 (J. Trudeau, *et al.*, 2016); and the EU has committed a more ambitious target of reducing 80 per cent greenhouse gas emission and a share of RES in electricity consumption reaching 97 per cent by 2050 (E. Commission, 2012). Moreover, some researchers look further ahead and propose even more ambitious goals, such as the future zero-carbon electrical grid of Australia (M. Wright, *et al.*, 2010)and 100 per cent renewables' scenarios (D. Crawford, et al., 2012). The power systems in most countries today are overstressed, overaged, and fragile, showing their inability to support the everlasting increase in electricity demand and incorporate the ambitions and trends of increasing renewable penetration. Such inadequacies in conventional power systems call for a series of transformations and updates in the power industry to address all the new challenges. At high voltage transmission level, the focus is mainly on the system's stability in an environment of higher renewable penetration, complicated demand side activities, and deregulated market competitions (T.V. Custem and C. Vournas, 1998; P. Kundur, *et al.*, 1994; C. Taylor, *et al.*, 1994).

Power system stability is defined as the ability of an electric power system for a given initial operating condition to regain a state of operating equilibrium after being subjected to a physical disturbance, with most system variables bounded so that practically the entire system remains intact (P. Kundur, *et al.*, 2004). The operation of a power system is inevitably exposed to various disturbances, such as an electric short-circuit on a transmission line and an unexpected outage of generator; when the disturbance is sufficiently severe, the power system can lose its stability.

This stability issue has been highlighted mainly due to its catastrophic consequences. Loss of power system stability can result in cascading failure and/or even widespread blackout events. Indeed, recent major blackouts over the world vividly demonstrate the far-reaching impacts of instability (T.V. Custem and C. Vournas, 1998). For instance, in 1996, the cascading disturbances in West Coast transmission system caused widespread blackout, which interrupted the electricity supply to 12 million customers for up to eight hours, and the cost of this event was estimated to be $2 billion (P. Kundur, *et al.*, 1994). Later, in 2003, a blackout in northeast affected 50 million customers and led to a $6 billion financial loss (T.V. Custem and C. Vournas, 1998). In 2012, two severe blackouts in India left over 600 million people in the dark for 10-12 hours (L.L. Lai, et al., 2013). More recently, in 2016, the high-wind-penetrated power system in South Australia failed to ride through six successive voltage disturbances caused by a storm, resulting in a statewide blackout and affecting over 1.67 million customers (A.E.M. Operator, 2016). To avoid the catastrophes as above, it is fundamental yet essential to maintain stable power system operation.

1.2 Mathematical Model

Generally, the power system dynamics can be modeled by a large set of differential algebraic equations (DAE) as follows:

$$\dot{\mathbf{x}} = \mathbf{f}(\mathbf{x}, \mathbf{y}, \mathbf{p}, \lambda)$$

(1.1)

$$0 = \mathbf{g}(\mathbf{x}, \mathbf{y}, \mathbf{p}, \lambda)$$

(1.2)

where set (1.1) corresponds to the differential equations of the system components, including generators, motors, dynamic loads, as well as their control systems, etc.; set (1.2) corresponds to the algebraic equations of the network and static loads. Vector **x** denotes the state variable (e.g. generator angles and speeds); vector **y** denotes algebraic variables (e.g. static load voltages and angles); vector **p** stands for controllable parameters (e.g. AVR set-points); and vector λ represents uncontrollable parameters (e.g. load levels).

It can be seen that power system dynamics is a strong non-linear high-dimensional complex problem. A straightforward yet essential approach for analyzing power system stability is time-domain simulation (TDS), which computes the system's dynamic trajectories along with other important system parameters by solving, step-by-step, equations (1.1) and (1.2), for a time-window of interest.

1.3 Classification

According to the physical nature of the resulting mode of instability, power system stability can be divided into three main categories (P. Kundur, *et al.*, 2004): rotor angle stability, voltage stability, and frequency stability. For each category, according to the size of the disturbance as well as the devices, processes, and the time span in assessing the stability, the stability can be further divided into *large-disturbance stability* or *small-disturbance stability* and *short-term stability* or *long-term stability*.

The classification of power system stability is illustrated in Fig. 1.1.

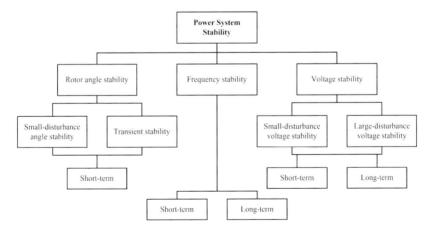

Fig. 1.1: Classification of power system stability (P. Kundur, *et al.*, 2004)

1.4 Rotor Angle Stability

Rotor angle stability refers to the ability of synchronous machines of an interconnected power system to remain in synchronism after being subjected to a disturbance. It depends on the ability to maintain/restore equilibrium between electromagnetic torque and mechanical torque of each synchronous machine in the system. Instability, that may result, occurs in the form of increasing angular swings of some generators, leading to their loss of synchronism with other generators. The outcome of the instability is the tripping of generators and separation of the whole system.

As shown in Fig. 1.1, the rotor angle stability can be divided further into large-disturbance rotor angle stability (also known as transient stability) and small-disturbance rotor angle stability, according to the size of the disturbance.

1.4.1 Large-disturbance Rotor Angle Stability

Large-disturbance rotor angle stability is concerned with the ability of the power system to maintain synchronism when subjected to a severe disturbance, such as a short circuit on a transmission line. The resulting system response involves large excursions of generator rotor angles and is influenced by the nonlinear power-angle relationship. Figures 1.2 and 1.3 show the simulated post-disturbance rotor angles of a real-world large power system for a stable case and an unstable condition (the two figures come from the simulations of a realistic large power system). As the figures show, the stability corresponds to the maintenance of synchronism while the instability corresponds to the loss of synchronism after the disturbance.

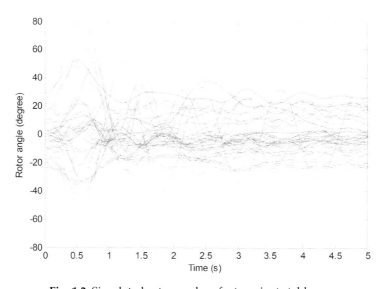

Fig. 1.2: Simulated rotor angles of a transient stable case

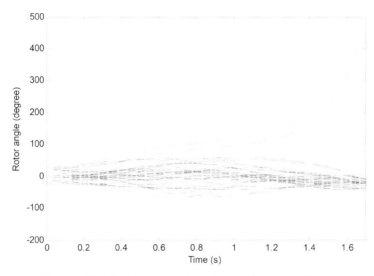

Fig. 1.3: Simulated rotor angles of a transient unstable case

1.4.2 Small-disturbance Rotor Angle Stability

Small-disturbance rotor angle stability is concerned with the ability of the power system to maintain synchronism under small disturbances. In contrast to transient stability, the disturbances are considered to be sufficiently small, e.g. the sudden variation of the load demands and/ or the output change of the generators, which can occur on a continuous basis. Small-disturbance stability depends on the initial operating state of the system. Instability usually appears as the rotor oscillations of increasing amplitude due to lack of sufficient damping torque, and the oscillations of power flow along transmission lines are accompanied. In today's power systems, small-disturbance rotor angle stability problem is usually caused by long-distance power transmissions.

A simulation example of the power flow oscillations of an interconnector due to the small-disturbance instability is shown in Fig. 1.4.

1.5 Voltage Stability

Voltage stability refers to the ability of a power system to maintain steady voltages in all buses in the system after being subjected to a disturbance from a given initial operating condition. It depends on the ability to maintain/restore equilibrium between load demand and load supply from the power system. Instability that may result occurs in the form of a progressive fall or rise of voltages of some buses. A possible outcome

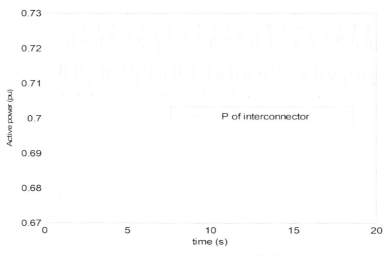

Fig. 1.4: Active power oscillation due to the small-disturbance instability

of voltage instability is loss of load in an area, or tripping of transmission lines and other elements by their protective systems, leading to cascading outages.

The driving force for voltage instability is usually the loads; in response to a disturbance, power consumed by the loads tends to be restored by the action of motor slip adjustment, distribution voltage regulators, tap-changing transformers, and thermostats. Restored loads increase the stress on the high voltage network by increasing the reactive power consumption and causing further voltage reduction. A run-down situation, causing voltage instability, occurs when load dynamics attempt to restore power consumption beyond the capability of the transmission network and the connected generation.

1.5.1 Large-disturbance Voltage Stability

Large-disturbance voltage stability refers to the system's ability to maintain steady voltages following large disturbances, such as system faults, loss of generation, or circuit contingencies. This ability is determined by the system and load characteristics, and the interactions of both continuous and discrete controls and protections.

The simulated post-disturbance voltage trajectories for voltage stability and instability are respectively shown in Figs. 1.5 and 1.6 (these two figures come from the simulation of the New England 10-machine 39-bus system) (A.M. Pai, 2012). For the stable case, the voltage trajectories swing within a tolerable bound and finally recover to the nominal level; while for the unstable case, the voltage trajectories oscillate drastically without reaching a stable level.

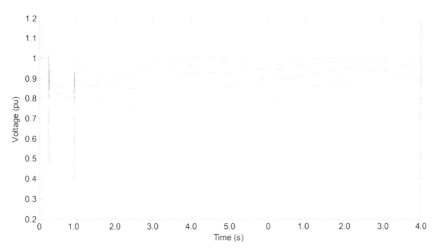

Fig. 1.5: Simulated voltage trajectories of the large-disturbance voltage stable case

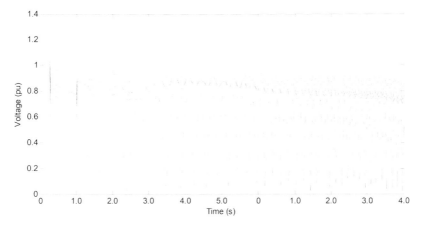

Fig. 1.6: Simulated voltage trajectories of the large-disturbance
voltage unstable case

1.5.2 Small-disturbance Voltage Stability

Small-disturbance voltage stability refers to the system's ability to maintain steady voltages when subjected to small perturbations, such as incremental changes in system load. In this case, the system voltage would drop steadily and continuously with the load increase until a critical point is reached; after the critical point, the system voltage would drop dramatically and finally collapse. A simulated power-voltage (PV) curve is shown in Fig. 1.7, where the loading parameter means the increasing rate of load. As it shows, after the load increases to 3.7115 times of the initial operating point (OP) the voltage instability occurs.

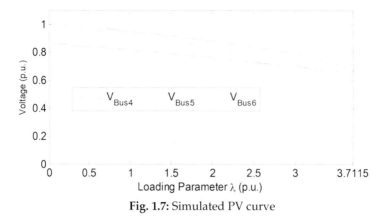

Fig. 1.7: Simulated PV curve

1.6 Frequency Stability

Frequency stability refers to the ability of a power system to maintain steady frequency following a severe system upset resulting in a significant imbalance between generation and load. It depends on the ability to maintain/restore equilibrium between system generation and load, with minimum unintentional loss of load. Instability, that may result, occurs in the form of sustained frequency swings, leading to tripping of generating units and/or loads.

Figures 1.8 and 1.9 show respectively the post-disturbance frequency trajectory of an unstable and a stable case. For frequency instability, the system frequency declines continuously to an unacceptable level. For the stability case, after the system frequency drops to a certain value, the automatic devices, such as under-frequency load shedding (ULFS) equipment, start to act and recover the system frequency to the nominal level.

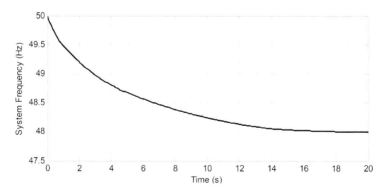

Fig. 1.8: Simulated post-disturbance frequency trajectory of a frequency unstable case

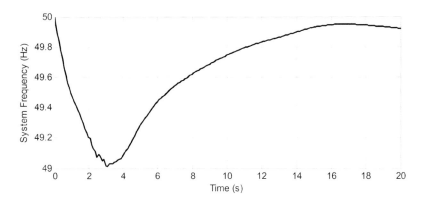

Fig. 1.9: Simulated post-disturbance frequency trajectory of a frequency stable case

1.7 Stability Criteria

Although power system stability is a complex problem, we can still build mathematical models to describe the problems by observing the explicit behavior of the system under different stability conditions. Such numerical models can provide criteria information to distinguish stable and unstable events as well as quantify the severity of the problem, which can facilitate power system stability analysis and guide stability control actions. The stability criteria are often presented in the form of stability indices or stability margins. This section introduces the common criteria used for transient stability, voltage stability, and frequency stability.

1.7.1 Transient Stability Index

Transient stability refers to the ability of synchronous machines of an interconnected power system to remain in synchronism after being subjected to a disturbance, and the instability usually appears in the form of increasing angular swings of some generators leading to their loss of synchronism with other generators.

Mathematically, the dynamic behavior of rotor angles can be described by a set of DAEs (also known as swing equations) and its classic model can be

$$\begin{cases} M_i \dfrac{d\omega_i}{dt} = P_{mi} - P_{ei} \\ \dfrac{d\delta_i}{dt} = \omega_i \end{cases} \quad \{i \in S_G\} \tag{1.3}$$

where for generator i, ω_i and δ_i are respectively the angle and angular speed; P_{mi} and P_{ei} are respectively the mechanical input and electrical output powers; M_i is the machine's inertia constant. Transient Stability Indices (TSI) are used to explicitly indicate the stability. Generally, they can be classified into the following categories:

1. Maximum rotor angle deviation during the transient period (D. Gan, *et al.*, 2000). The following transient stability degree is usually calculated as the TSI

$$\text{TSI} = \frac{360 - \left|\Delta\delta_{\text{max}}\right|}{360 + \left|\Delta\delta_{\text{max}}\right|} \times 100 \tag{1.4}$$

   where $\Delta\delta_{max}$ is the absolute value of the maximum angle deviation between any two generators during the TDS. If this TSI > 0, the rotor angle is defined as 'stable', and otherwise 'unstable'.
2. Transient energy, which is determined by the kinetic energy and potential energy of a post-fault power system (M. Pai, 2012).
3. Swing margin η, which is derived by applying equal-area criterion (EAC) to an one-machine infinite bus (OMIB) system about the groups of critical machines (CMs) and non-critical machines (NMs) that are defined by extended equal-area criterion (EEAC) theory, including IEEAC and single-machine equivalent (SIME) approaches (D. Ruiz-Vega and M Pavella, 2003; A. Pizano-Martianez, *et al.*, 2009, R. Zarate-Minano, *et al.*, 2009 and A. Pizano-Martinez, *et al.*, 2011). By observing the behavior of mechanical and electrical power of the OMIB in the P-δ plane, the margin η is defined as follows:

$$\eta = A_{\text{dec}} - A_{\text{acc}} \tag{1.5}$$

   where δ is the rotor angle; A_{dec} and A_{acc} refer to the decelerating and accelerating area, respectively; a positive η means transient stable and a negative η means unstable. The details of EEAC theory can be found in (Y. Xue, *et al.*, 1989). and some applications of this criterion are available in (Y. Xu, *et al.*, 2012 and Y. Xu, *et al.*, 2017).

The first TSI is basic and general but it provides very little information about the stability degree; besides, it needs a heuristic threshold to verdict the stability that is usually system dependent and not easy to define. The second one is based on the transient energy function (TEF) approach (M. Pai, 2012) and can be a quantitative measurement of the stability degree, but it commonly suffers from conservative property. The last one is derived from the whole system trajectories from full TDS, and can withhold the multi-machine dynamics; besides, the byproducts in calculating the stability margin, such as CMs and NMs, time to instability, and time to first-swing stability are all valuable for transient stability analysis (M. Pavella, *et al.*, 2012 and Y. Xue, *et al.*, 1989).

Besides, the critical clearing time (CCT), which is the maximum acceptable fault clearing time without losing stability, can also be viewed as a TSI, but it relies on a dependent TSI (such as the above ones) to define the 'stability'.

1.7.2 Voltage Stability Index

The concerns on voltage stability can be manifested from different aspects, such as the size of disturbance, the length of interested time period, etc. In the literature, the voltage stability indices are broadly designed for two voltage-stability problems: static voltage stability and short-term voltage stability (STVS). The conventional and recently developed indices for these two voltage stability problems will be introduced.

1.7.2.1 Static Voltage Stability Index

Static voltage stability is related to the existence of the operating equilibrium of the system after or without a contingency. Therefore, the analysis mainly focuses on evaluating the proximity of the system to instability.

There are a number of indicators proposed to quantify the proximity or 'distance' to static voltage instability. The most commonly used one is the loadability margin (LM), which measures how much power can still be delivered to the loads at the base OP along a certain loading incremental direction before the voltage collapses. The LM criterion has been widely used in previous works (S. Gerbex, *et al.*, 2001; N. Yorino, *et al.*, 2003, Y.-C. Chang, 2011; E. Ghahramani and I. Kamwa, 2012). However, to calculate LM, it is necessary to define a probable loading pattern. While such a pattern can be determined in the on-line operation or operational planning stage, it is very difficult to be predicted in the long run. More importantly, the LM can be quite sensitive to the variations in the loading pattern.

Another system-dependent only index, called voltage collapse proximity indicator (VCPI), is also widely used for quantifying the static voltage stability degree. VCPI was originally proposed in (M. Moghavvemi and O. Faruque, 1998) and its accuracy and reliability have been rigorously demonstrated therein. As a line-based index, it involves both power and voltage variation and can provide reliable information about the proximity to voltage collapse. According to recent comparison studies (C. Reis and F.M. Barbosa, 2006; M. Cupelli, *et al.* 2012), VCPI has the best accuracy and robustness performance among others. Other attractive advantages of VCPI are simplicity, high calculation speed, and flexibility for simulating any type of topological and load modifications.

VCPI is based on the concept of maximum power transferred through the lines of a network as shown in Fig. 1.10.

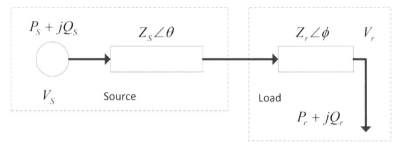

Fig. 1.10: Single transmission line model

VCPI is mathematically calculated as:

$$VCPI\,(power) = \frac{P_r}{P_{r(\max)}} = \frac{Q_r}{Q_{r(\max)}} \tag{1.6}$$

where P_r and Q_r are the real and reactive power transferred to the receiving end, respectively, and can be obtained from the power flow calculation. $P_{r(\max)}$ and $Q_{r(\max)}$ are the maximum real and reactive power that can be transferred through a line, respectively, which can be calculated as follows:

$$P_{r(\max)} = \frac{V_S^2}{Z_S} \frac{\cos\phi}{4\cos^2\left((\theta-\phi)/2\right)} \tag{1.7}$$

$$Q_{r(\max)} = \frac{V_S^2}{Z_S} \frac{\sin\phi}{4\cos^2\left((\theta-\phi)/2\right)} \tag{1.8}$$

where V_s is the voltage magnitude of the sending end, Z_s is the line impedance, θ is the line impedance angle, and $\phi = \tan^{-1}(Q_r/P_r)$ is the phase angle of the load impedance.

The value of *VCPI* (*power*) ranges from 0 to 1, where a larger value indicates a shorter distance to voltage collapse and, therefore, a lower degree of static voltage stability. It is also straightforward to identify critical lines. In practice, to evaluate the overall level of the voltage stability, it is useful to use the sum of the *VCPI* (*power*) of all transmission lines in the system.

$$VCPI^T = \sum_{l=1}^{L} VCPI\left(power\right)_l \tag{1.9}$$

where L is the total number of transmission lines in the system.

1.7.2.2 STVS Index

To evaluate the short-term voltage performance, some industrial criteria, such as NERC criteria (D. Shoup, *et al.*, 2004) have been established. These criteria are defined on the basis of unacceptable voltage dip magnitude and the duration time, but they can only provide a binary answer to the voltage stability, namely, stable or unstable.

Given the fast variation of post-disturbance voltage trajectories, STVS analysis mainly focuses on fast voltage collapse and unacceptable transient voltage deviation (i.e. fault-induced delayed voltage recovery (FIDVR)) following a disturbance. The detailed explanation on different STVS phenomena is further provided in Section 6.4.1. The evaluation of transient voltage collapse is a binary question, that is, whether or not the system loses the short-term equilibrium following a disturbance. Consequently, the index to detect fast voltage collapse, namely transient voltage collapse index (TVCI), can be a binary index, namely, 1 or 0.

To quantitatively measure the degree of transient voltage deviation, several indices have been proposed in the literature. A. Tiwari and V. Ajjarapu (2011) proposed a *contingency severity index* which is the sum of two indices respectively accounting for the voltage magnitude violation and the duration time. However, it only considers the maximum voltage deviation magnitude. Y. Xue, *et al.* 1999 proposed an index called *transient voltage dip acceptability* to consider both the voltage dip magnitude and its duration time, but a non-linear curve-fitting process is needed during its calculation. Y. Dong, *et al.* 2015 proposed a *voltage sag severity index* to evaluate the voltage recovery performance on each single bus, but it lacks systematic evaluation on the stability degree.

An alternative yet more straightforward index, called *transient voltage severity index* (TVSI), was also developed in our previous work (Y. Xu *et al.*, 2014) to quantify the transient voltage performance in a systematic view:

$$TVSI = \frac{\sum_{i=1}^{N} \sum_{t=T_c}^{T} TVDI_{i,t}}{N \times (T - T_c)} \tag{1.10}$$

where N is the total number of buses in the system, T is the considered transient time frame, T_c is the fault clearing time, and TVDI is the transient voltage deviation index, calculated by:

$$TVDI_{i,t} = \begin{cases} \dfrac{|V_{i,t} - V_{i,0}|}{V_{i,0}}, & \text{if } \dfrac{|V_{i,t} - V_{i,0}|}{V_{i,0}} \geq \mu \\ 0, & \text{otherwise} \end{cases} \quad \forall t \in [T_c, T] \tag{1.11}$$

where $V_{i,t}$ denotes the voltage magnitude of bus i at time t, which is obtained from TDS and μ is the threshold to define unacceptable voltage deviation level, which can be set according to the industrial criteria, e.g. 20 per cent.

The concept of TVSI is illustrated in Fig. 1.11. TVSI only accounts for the buses with unacceptable voltage violation during the transient

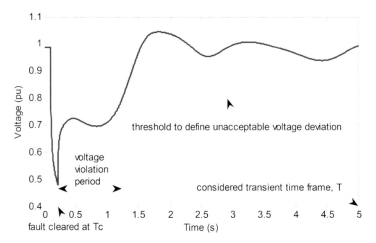

Fig. 1.11: Illustration of the concept of TVSI

period, and it measures not only the voltage violation magnitude, but also the associated duration time in the whole system level. It can provide a quantitative comparison of the system's transient voltage performance following a disturbance. A smaller TVSI value means that the transient voltage performance is better.

Although TVSI can systematically evaluate the voltage deviation severity, such direct averaging operation in TVSI calculation can hardly reveal the voltage deviation on a single bus, if all the other buses remain steady following the disturbance. This may lead to biased evaluation on the STVS event.

Considering the above inadequacies, the TVSI can be upgraded into a new index, called *root-mean-squared voltage-dip severity index* (RVSI), to evaluate the voltage deviation severity:

$$RVSI = \sqrt{\frac{\sum_{i=1}^{N} VSI_i^2}{N}} \tag{1.12}$$

where VSI is short for *voltage-dip severity index*, which quantifies the unacceptable voltage deviation of each single bus

$$VSI_i = \int_{T_c}^{T} D_{i,t} dt \tag{1.13}$$

where

$$D_{i,t} = \begin{cases} \dfrac{V_{i,0} - V_{i,t}}{V_{i,0}}, & \text{if } V_{i,t} \leq (1-\mu)V_{i,0} \\ 0, & \text{otherwise} \end{cases} \quad \forall t \in [T_c, T] \tag{1.14}$$

The concept of RVSI is illustrated in Fig. 1.12. The VSI quantifies the FIDVR severity of every single bus, and is computed as the area covered by the part of voltage trajectory that is below the threshold level. Such area quantity is scaled by both voltage-dip magnitude and duration, meaning VSI complies with the design concept of the industrial criteria (D. Shoupm *et al.* 2004). RVSI is the root mean squared average of the VSI values of all buses. Compared to the arithmetic mean, the advantage of root mean square is its ability of emphasizing the buses with higher VSI value (i.e. slower voltage recovery) in the averaging process. In doing so, even if most buses in the system undergo fast voltage recovery and only one or two buses undergo much slower recovery, such FIDVR severity can still be fairly reflected in the RVSI value.

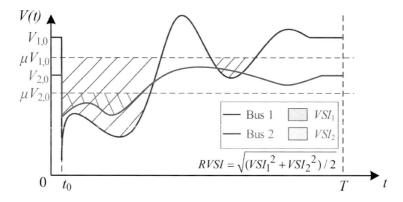

Fig. 1.12: Graphical illustration of RVSI

1.7.3 Frequency Stability Index

Power systems are required to operate within a normal frequency level, such as 50 Hz ± 0.2 Hz. When subjected to a disturbance, the frequency will fluctuate and when the disturbance is sufficiently severe, the frequency may continuously decline or rise to an extent that may harm the power plants, the load-side equipment, and even the system integrity. It depends on the ability to maintain/restore equilibrium between system generation and load, with a minimum unintentional loss of load. Instability that may result occurs in the form of sustained frequency swings, leading to tripping of generating units and/or loads.

Traditionally, the frequency stability assessment (FSA) is conducted by TDS, which is to numerically simulate the frequency response after a contingency (this involves solving a large set of DAEs that represent the system and the fault) and observe if there is any violation of frequency requirement.

In order to quantify the frequency stability, the frequency stability margin (FSM) was proposed in (X. Taishan and X. Yusheng, 2002). Let a set of two-element tables $[(F_{cr,1}, T_{cr,1}),\ldots, (F_{cr,i}, T_{cr,i})]$ describe the frequency dip acceptability for each bus. The frequency stability can be determined accordingly if the maximum duration time that the bus frequency below $F_{cr,i}$ is smaller than the corresponding $T_{cr,i}$.

The FSM can be calculated by:

$$\eta = \begin{cases} T_{cr,i} - T_i & , \ F_{\min,i} \leq F_{cr,i} \\ F_{\min,i} - F_{cr,i} + T_{cr,i} & , \ F_{\min,i} > F_{cr,i} \end{cases} \tag{1.15}$$

where $F_{\min,i}$ is the minimum frequency of the bus i during the transient period.

In order to reduce the two-dimension frequency stability requirement $(F_{cr,i}, T_{cr,i})$ into one-dimension so as to decrease the computation burden of the FSM, a converting factor k is employed, and $(F_{cr,i}, T_{cr,i})$ becomes $(F'_{cr,i}, 0)$, where $F'_{cr,i} = F_{cr,i} - k \cdot T_{cr,i}$. And the FSM becomes:

$$\eta = \left(F_{\min,i} - F'_{cr,i} \right) \times 100\% \tag{1.16}$$

The FSM ranges from −100 to +100, when the FSM is larger than zero. It means the system is secure; otherwise, the system is insecure. The concept of FSM is very valuable since it can numerically measure the stability degree; even more importantly, it opens avenues to sensitivity approach, based on which, the frequency stability control can be systematically designed on an optimization basis (X. Taishan and X. Yusheng, 2002).

1.8 Conclusion

The power system stability problem has been highly concerned in today's environment of higher renewable penetration, complicated demand-side activities, and deregulated market competition. This chapter defines power system stability, presents the mathematical model, and summarizes the classification strategies on different power system stability phenomena.

Power system stability is defined as the ability of an electric power system for a given initial operating condition to regain a state of operating equilibrium after being subjected to a physical disturbance, with most system variables bounded so that practically the entire system remains intact. It depends on power system dynamics, which is a strong non-linear high-dimensional complex problem that can be mathematically modeled by a large set of DAEs.

The classification of power system stability can be manifold. Firstly, according to the physical nature of the resulting mode of instability, power system stability can be divided into rotor angle stability, voltage stability, and frequency stability. Secondly, according to the size of the concerned disturbance, the stability can be further divided into large-disturbance stability and small-disturbance stability. Thirdly, according to the time span in assessing the stability, the stability problems can be divided into short-term stability and long-term stability problems.

To numerically distinguish stable and unstable events and/or evaluate the severity of the events, numerical stability indices are needed for the different stability phenomena. The three commonly used TSIs include maximum rotor angle deviation, energy function, swing margin, and CCT. The voltage stability indices are designed separately for static voltage stability and STVS problems. The numerical index for static voltage stability is VCPI, whereas the voltage deviation in STVS can be evaluated using TVSI and RVSI. The main stability index for frequency stability is FSM.

Stability Assessment and Control: Problem Descriptions and Classifications

2.1 Introduction

Power system stability is concerned with the system's dynamic behavior following a physical disturbance. The loss of stability can lead to catastrophic consequences, such as cascading failure and even widespread blackout. Maintaining power system stability has long been an essential requirement for secure and continuous electricity supply to the customers.

With the intermittent generation from RES and the active demand-side participation, system operation is exposed to more variations. In such a highly-volatile operating environment, the conventional power system stability analysis scheme, that is implemented in an off-line manner, is considered not fast enough to cope with the ongoing variations, intermittencies, and uncertainties. With the advancement in power system monitoring technologies, on-line and real-time power system stability analysis can be achieved and is of great significance in supporting power system stability.

Wide-area measurement systems are envisaged as the next-generation grid sensing and communication infrastructure for enhanced situational awareness. Based on the massive amount of data from time-synchronized measurement devices, such as PMUs, data-driven methods have been identified as powerful tools for on-line and real-time stability assessment and control (SA&C) given the high complexity of the system and difficulties in modeling the physics behind the complex system dynamics.

This chapter provides an overview of the SA&C problem in power system operation, which reviews the conventional off-line stability analysis, presents the advanced stability monitoring infrastructure, introduces the on-line and real-time stability analysis, classifies the different SA&C schemes, and discusses the stability analysis requirements for today's power systems.

2.2 Off-line Stability Analysis

Conventionally, the power system stability is assessed based on off-line study which is performed at an appropriate lead time, e.g. day ahead. Based on the forecasted operating conditions, a large number of possible disturbances are screened at off-line stage to find the insecure disturbances and to identify the situations on which the operators should concentrate. During on-line power system operation, the off-line analysis results are used by the operators in a look-up manner.

TDS is the conventional method adopted for power system stability analysis. It simulates the system dynamic trajectories by iteratively solving a set of different algebraic equations (DAE) that are presented in (1.1) and (1.2). TDS shows some strength in power system stability analysis (M. Pavella, *et al.*, 2012). First, it provides essential information about the system's dynamic evolution (e.g. generator swing curves, bus voltage trajectories, system frequency trajectory) following the disturbance. Second, it is adaptable to any power system modeling and fault scenario. Third, it achieves sufficient accuracy in its result, provided the modeling of the system is available and the system parameters are accurately known.

However, since TDS needs an iterative calculation to solve the large set of DAEs, it suffers from high computation burden. For a real-world power system with a large number of buses, it takes TDS several hours to complete its simulations on all the postulated disturbances. With such high computation burden, TDS is mostly used in off-line stability study.

2.3 Advanced Power System Monitoring Infrastructure

The supervisory control and data acquisition (SCADA) systems have served as the power system monitoring infrastructure for more than a decade. It has significantly improved the power system sensing and enabled transmission automation. Essential electrical quantities, such as voltage, current and power flow of all the key nodes in the electricity grid are scanned every few seconds. This information to some extent has facilitated power system security and stability analysis. However, the SCADA systems lack the ability to provide measurements on phase-angle differences across the interconnected power system.

The recent widespread use of the global positioning system (GPS) has made time synchronization and phase-angle measurements available. PMU is an electronic device that takes measurement of voltage/current phasors on a power grid, using a common time reference from the GPS. The hardware block diagram of a PMU is shown in Fig. 2.1 (A.G. Phadke and J.S. Thorp, 2008). Basically, the PMU converts the analogue voltage/

current waveforms measured by potential/current transformers into digital measurement data. Adhering to the GPS time reference with an accuracy of 1 microsecond, the phasor-locked oscillator ensures that all the phasor measurements over the power grid are time-synchronized, meaning the whole grid is measured in a system-wide manner. Moreover, the measurement output rate of a PMU can reach up to 60 samples per second. With such high temporal resolution, PMUs are devices that facilitate taking real-time measurements.

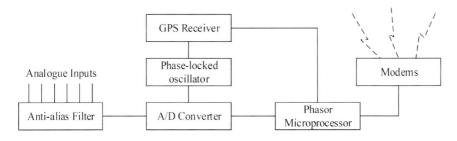

Fig. 2.1: Hardware block diagram of PMU

WAMS takes advantage of the time-synchronized measurement ability of PMUs to allow real-time power system sensing over large areas. In a WAMS, the measurement data from PMUs are collected and processed under a hierarchical structure as shown in Fig. 2.2 (A.G. Phadke and J.S. Thorp, 2008). The data provided by the PMUs in an area of the system are collected by phasor data concentrators (PDC), and then transmitted to the control centre via high-speed data communication networks. Compared to the existing SCADA system based on static-state estimation, WAMS, as a prospective monitoring infrastructure, shows the following advantages:

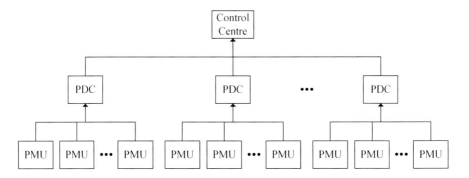

Fig. 2.2: WAMS structure

- The synchronization of phasor measurements provides wide-area observability on power grids, which further facilitates system-wide coordinated power grid analysis.
- The high measurement speed (i.e. up to 60 samples per second) enables real-time visibility of power grids, thereby capture of the dynamic behavior of power grids becomes possible.

2.4 On-line and Real-time Stability Analysis

The new features of today's power systems have significantly complicated the power system operation. In the new environment, the off-line usually analysis becomes inadequate due to the following reasons:

- The improvement in energy-use efficiency requires the system to transfer more power, but the conventional off-line paradigm is always conservative in determining the available transfer capability.
- The intermittent RES and the active demand-side participation bring more uncertainty to power system operation from time to time. Although a large amount of effort has been put to improve the forecast performance on RES generation (C. Wan, *et al.*, 2014) and load demand (W.C. Kong, *et al.*, 2019), the off-line stability analysis based on day-ahead forecast can hardly capture the real-time operating conditions of the system in a highly volatile environment.

Benefiting from the advanced sensing and communication technologies, on-line and real-time stability analysis is becoming possible to maintain system stability and eliminate blackout risk. On-line and real-time stability analysis is able to monitor and evaluate the system's stability condition on a continuous basis, and timely triggers the stability control actions, whenever necessary.

In on-line and real-time stability analyses, the stability assessment (SA) results must be refreshed based on each measurement snapshot of the system, which requires fast computational speed to provide instantaneous response. Once an unstable event is detected via SA, control actions must be taken to regain power system stability. However, the conventional TDS is very time consuming so that it restricts its applications in on-line or real-time stability analysis. Therefore, more advanced methodologies with faster decision-making capability are imperatively needed to satisfy on-line and real-time needs.

2.5 Classification of Stability Assessment and Control

In sophisticated power system operations, the on-line and real-time stability analyses can be classified into two different modes to avoid blackout events. The proceeding of the two modes is shown in Fig. 2.3.

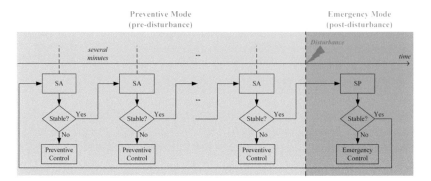

Fig. 2.3: Classification of power system SA&C

- *Preventive mode*: The power system's stability analysis in preventive mode is also called SA, which is carried out every several minutes when the system is operating at its steady state. SA aims at screening a number of disturbances with a high probability of occurrence to examine the ability of current operating condition of the system to withstand the disturbances. Different stability criteria, including rotor-angle, voltage, and frequency stability, should all be examined. Following any of the postulated disturbances, if the system is able to survive the dynamic transition and regain an acceptable operating equilibrium, the current operating condition should be assessed as stable. If the system is assessed as unstable, preventive control actions, such as generation rescheduling, will be triggered to force the system to operate under a stable condition.
- *Emergency mode*: In emergency mode, since the system can lose its stability within a very short timeframe (e.g. several seconds) following the disturbance, the stability status of the system must be predicted immediately after experiencing the disturbance. In this sense, the power system's stability analysis in emergency mode refers to the stability prediction (SP). This SP means deciding whether the system is in the process of losing its stability in response to the ongoing disturbance. Once an unstable propagation is predicted, the emergency control actions, such as load shedding, generation tripping, VAR control, and network separation, need to be activated to regain stable system operation. The stability status should be predicted as early as possible to timely trigger the emergency control actions and avoid further cascading failure and/or blackout events.

2.6 Requirements for Stability Analysis Tools

Conventionally, power system stability is analyzed in an off-line manner. Such off-line approach can be quite sufficient to maintain system stability in a period when power industry has been vertically integrated and regulated. However, the recent changes from conventional power grid into smart grid, on one hand, have improved the efficiency of power industry with both economy and environmental merits; on the other hand, they have complicated the system operation conditions, rendering the conventional off-line approaches inadequate and non-economical. The main challenges in smart grid faced by conventional stability analysis approaches can be summarized as follows:

- In context of power market, the economy becomes one of the major concerns, which requires the system to transfer more electricity. However, traditional off-line method is always conservative in determining the available transfer capability.
- Meanwhile, the infrastructure investment is unmatched with the rapid electricity load growth, which has pushed the system to operate closer to its stability boundary.
- Furthermore, the market activities can drive the system into an erratic power transfer pattern, which is difficult to forecast.
- With increased RES (especially wind and solar), the system becomes furthermore unpredictable due to the inherent intermittency of these energies. Off-line methods thereby become inapplicable to capture realistic operating conditions in the highly uncertain environment.

The conventional TDS is computational burdensome, which restricts its applications in on-line or real-time stability analysis. Thus more advanced computational tools for power system stability analysis are imperatively needed to satisfy the on-line and real-time analyses needs. Considering the above issues in the new environment, the new tools are expected to satisfy the following requirements:

- Higher accuracy to allow more transfer capability with the existing resources.
- Higher efficiency (e.g. on-line and/or real-time analysis capability) to capture the variations and uncertainties in smart grid.
- Ability to extract reliable stability control rules.
- Higher tolerance to measurement system events, such as incomplete measurements, communication delay, and measurement noise, etc.

2.7 Conclusion

This chapter first reviews the conventional off-line stability analysis and demonstrates its inadequacy for today's power systems. A brief introduction on the advanced power system monitoring infrastructure, including PMU and WAMS, are also provided. Based on such advanced grid sensing, on-line and real-time power system stability analyses could be realized to deal with the challenges posed by energy efficiency requirement, RES integration, active demand side management, etc.

This chapter further classifies on-line and real-time stability analyses into two modes: preventive mode and emergency mode. In preventive mode, SA is carried out every several minutes when the system is operating at its steady state. SA screens a number of disturbances with a high probability of occurrence to examine the ability of current operating condition of the system to withstand the disturbances. In emergency mode, SP actions are taken after a disturbance actually occurs in the system. Since the system can lose its stability within a very short timeframe (e.g. several seconds) following the disturbance, real-time response speed is required for SP and emergency control.

The transformation from traditional grid into smart grid poses new challenges to maintain power system stability, which requires higher performance and more capability for future stability analysis tools, including higher accuracy, higher efficiency, ability to extract control rules, and higher tolerance to measurement system events.

CHAPTER
3

Intelligent System-based Stability Analysis Framework

3.1 Introduction

Electric power systems, the most complex manufactured systems in the world, maintain the secure and economical generation, transmission, and distribution of electricity. Thus, because of its complexity and importance, in recent years, there has been considerable research concerning how intelligent systems (IS) techniques can improve applications that manage key power systems, such as electricity market simulation, market risk management, power grid planning, and voltage control (C. Ramos and C.C. Liu, 2011; Z. Vale, *et al.*, 2011; Y.F. Guan and M. Kezunovic, 2011; J. Ferreira, *et al.*, 2011; C.F.M. Almeida and N. Kagan, 2011). Power system operations are routinely disrupted by both large and small disturbances. In fact, structural changes in the network, such as an electrical short-circuit or an unanticipated system-component loss, often cause large disturbances. Even small disturbances, such as continuously occurring small-load demand variations, can cause the power system to lose stability, triggering cascading failures or catastrophic blackouts. With the deployment of WAMS, monitoring power grids in a wide-area and time-synchronized manner can capture the dynamic behavior of the system. Based on such measurements, IS techniques have been recently identified as powerful tools for on-line and real-time power system stability analyses, given the high complexity of the system and difficulties in modeling the physics behind the complex system dynamics (Z.Y. Dong, *et al.*, 2013).

This chapter presents the preliminaries of IS strategy application to power system stability analysis, systematically describing the principles, advantages, and process of constructing an IS for power system SA&C, and discusses some most critical issues in its development and implementation stages. The perspectives on possible solutions to the problems are also discussed. The framework elaborated in this chapter serves as the baseline of IS methods.

3.2 Principles

Conventional methods for power system stability analysis are mainly based on TDS, which usually suffer from excessive computational burden, and inability to offer useful information about system stability characteristics and guideline for controls.

The advent of IS technologies provides an alternative and promising methodology for much faster and information-rich stability analysis. Generally, the IS is off-line constructed; when applied on-line, the computation time required to obtain system stability information can be dramatically reduced by eliminating the need for calculating any non-linear equations.

Unlike analytical methods that are based on mathematical calculations of system DAE, an IS can capture the relationship between power system operating states and the dynamic security status, by extracting comprehensive knowledge from a stability database. It only requires a system snapshot as input to determine the corresponding stability conditions. As soon as the input is fed, the SA&C results can be given within very short delays (typically not exceeding a couple of seconds), and the results can be utilized in real-time, combined with other extracted knowledge for decision support. A typical structure of such an IS is described in Fig. 3.1.

Compared with other state-of-the-art methods, the advantages of IS scheme include the following aspects:

- *Much faster speed*: Real-time stability status must, of course, be evaluated at the current or immediate future point. The IS needs only a fraction of a second to determine the system stability conditions after the inputs are fed, which thereby can offer much more response time to take remedial control actions, if necessary. Such high computational speed is usually sufficient to satisfy the on-line and real-time application needs.
- *Knowledge discovery*: IS is constructed by automatic learning from a database in the off-line training process. It can discover and extract useful information on system stability characteristics, which are valuable for better physical understanding of the power system and developing stability control strategies. In some cases, such knowledge can be described as interpretable rules, not only for guiding the control actions, but also for post-event auditing and forensic purposes.
- *Less data requirement*: Analytical methods strictly need accurate and full descriptions of power system, including power flow and system model parameters. However, this information can be uncertain or even unavailable in real-time environment. In contrast, the IS only requires significant and/or available input parameters to perform SA&C.

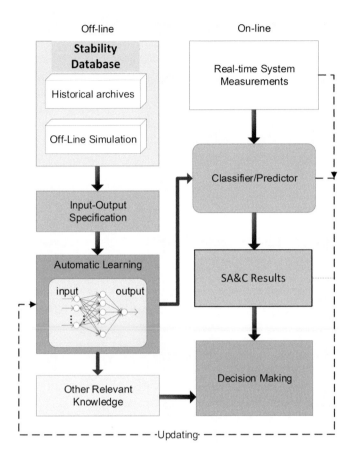

Fig. 3.1: Typical structure of an intelligent SA&C system

- *Strong generalization capacity*: Rather than exhaustively performing TDS on each pre-assumed scenario, the IS can, at one time, handle a wide range of stability analysis cases, previously seen and unseen.
- *Versatility*: IS can be a unified method to SA&C. Not to mention its adaptability in all the three types of stability categories, it can be designed to predict stability status (classification) and/or measure stability margin (regression), as well as to provide rules for stability controls.
- *Complementary to other tools*: In addition to being applied individually, the IS can also be integrated with conventional SA&C engines, working for contingency filtering, ranking and selecting to increase the overall SA&C computation efficiency.

The earliest systematic attempt of using artificial intelligence (AI) for fast SA was carried out in late 1980s (L. Wehenkel, *et al.*, 1989). Since then, research outputs on this topic have been increasing steadily (L. Wehenkel, *et al.*, 1994; C.A. Jensen, *et al.*, 2001; E.M. Voumvoulakis, *et al.*, 2006; K. Sun, *et al.*, 2007; Y. Xu, *et al.*, 2010; J. Zhao, *et al.*, 2007; Y. Kamwa, *et al.*, 2001; Y. Kamwa, *et al.*, 2010; Y. Kamwa, *et al.*, 2009; N. Amjady and F. Majedi, 2007; N. Amjady and S. Banihashemi, 2010; A.D. Rajapkse, *et al*, 2009). In the following chapter, we identify some most critical factors covering the development and implementation of an IS for SA&C, which are described in Fig. 3.2. The following chapter discusses these factors by answering questions involved at each stage. The objectives are to draw a clear picture about the process, difficulties, bottlenecks, and provide possible solutions.

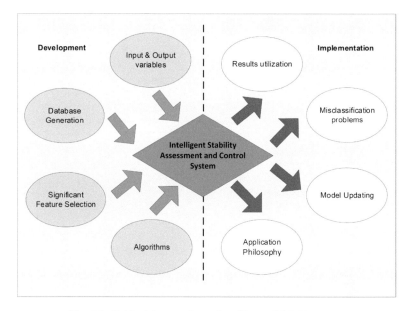

Fig. 3.2: Critical factors in an intelligent SA&C system

3.3 Development Stage

In the development stage, we may encounter the following critical questions:

- How to generate the stability database?
- Which kind of system variables to be used as input and output of the IS?
- Is there any need to select only some interesting features as input? And how?
- Which algorithm should be used to extract the knowledge?

3.3.1 Stability Database

A sufficient stability database is paramount to the IS because, if the database is biased, unrealistic, or too small, the extracted knowledge will probably be useless. Typically, generation of an adequate database consists of simulating a wide range of prospective system's operating conditions and acquiring their stability information. Historical operational archival data can also be included in the database.

However, it can be very difficult to prepare sufficient prospective system's operating conditions due to the uncertainties in market activities and non-conventional energies. A conservative way is to produce as many as possible OPs covering broad changes in load/generation pattern, voltage level, and network topology, etc. The uncertainties are considered by adding randomness or, more systematically, using *Monte-Carlo* simulation in the process. Standard TDS tools can be used to acquire the stability information of the produced OPs. In order to avoid over-fitting, a validation of the database should be conducted subsequently.

3.3.2 Input and Output

The variables characterizing the power system can fall into two categories: static and dynamic, respectively corresponding to pre- and post-fault environments. Some basic variables are given in Table 3.1, which have been commonly adopted in the literature (L. Wehenkel, *et al.*, 1989; L. Wehenkel, *et al.*, 1994; C.A. Jensen, *et al.*, 2001; E.M. Voumvoulakis, *et al.*, 2006; K. Sun, *et al.*, 2007; Y. Xu, *et al.*, 2010; J. Zhao, *et al.*, 2007; I. Kamwa, *et al.*, 2001; I. Kamwa, *et al.*, 2010; I. Kamwa, *et al.*, 2009; N. Amjady and S.F Majedi, 2007; N. Amjady and S. Banihashemi, 2010; A.D. Rajapakse, *et al.*, 2009).

Table 3.1: Basic power system variables

Static features	Dynamic features
P, Q load demand	Rotor angle
P, Q generation output	Rotor angular velocity
Voltage magnitude and angle	Rotor acceleration speed
P, Q line flow	Dynamic voltage trajectories
…	…

Both kinds of variables can be used as input to the IS, but they can lead to considerable differences, specifically:

With *pre-disturbance variables*, stability can be evaluated before contingency comes. If the current operating state is found unstable, operators can arm *preventive controls*, e.g. generation rescheduling to move the current OP into secure regions so as to avoid the risk of instability.

This ensures that the power system operates in a preventive state against anticipated contingencies. However, it would mean expensive due to the reallocation of power generation among generators.

By contrast, when using *post-disturbance variables* as input, the SA can only be performed after contingency really occurs; and if stability is shown to be getting lost, *emergency controls* should be instantly activated, e.g. generator tripping and/or load shedding. This counter measure brings no pre-paid cost, but will lead to tremendous economic and social expense if it is really invoked. Besides, the use of post-disturbance variables requires a certain response time, which has been shown ranging from 3 s to 1 s after the fault is cleared (I. Kamwa, *et al.*, 2001; I. Kamwa, *et al.*, 2010; I. Kamwa *et al.*, 2009; N. Amjady and S.F. Majedi, 2007; N. Amjady and S. Banihashemi, 2010; A.D. Rajapakse, et al., 2009). This waiting time can be too long for operators to take timely remedial action to stop the very fast instability development (loss of synchronism may arise within 1.5 s in some circumstances).

Another important factor regarding the input is the dependency on network topologies. The topology of a power system refers to the electrical connection status of the components to the grid. Such structural changes occur on a daily base because of regular maintenance and switching activities. When substantial topology changes happen, the previously generated database may not contain the information of the new system. Generally, pre-disturbance variables are more dependent than post-disturbance variables on network topology since they are static parameters.

The output of the IS, on the other hand, is the parametric indication of system's stability condition. It can be classified into nominal and numerical values, which determine the SA&C as a classification or regression problem. The former can be discrete labels, such as stable/unstable; the latter is continuous value, such as the stability margin. Besides, when strategically coded, the output can also represent control actions, e.g. generator rescheduling or load shedding used as a preventive or emergency control.

3.3.3 Significant Features

The inherent high-dimensional property of power system entails thousands of features for characterizing its operating state. Is it necessary to enclose all the features as input to an IS?

Indeed, it looks better if we use them all since this can wholly describe a system state. However, in practice, it will be beneficial to select only important variables (named significant features) as input for the following reasons:

- The use of significant features can drastically reduce the overall training time and thus improve the computation efficiency.

- Significant features can result in a higher SA accuracy because of the perpetual existence of redundant and noisy features relating to system stability.
- Significant features can provide other knowledge, such as system's critical/weak point.

It is worth mentioning that the last point is very valuable for developing stability control strategies. To illustrate this, an example is shown in Fig. 3.3 where OPs with different stability statuses are plotted in two two-dimensional distance space (extracted from [44]). In the upper window, the OPs are represented by significant features and are clearly

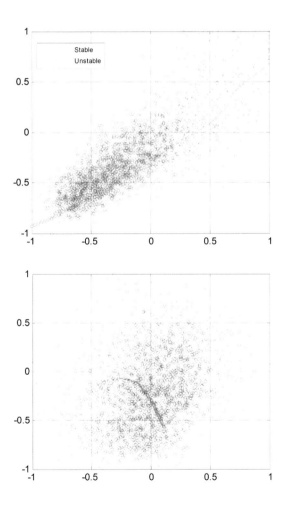

Fig. 3.3: Representing OP with significant and non-significant features

well separated, while in the lower window, where a non-significant features space is shown, the OPs are heavily overlapped.

It is obvious that the upper space provides a transparent relationship between the operating parameters and their stability status, and the knowledge on stability regions can be recognized. If one can tune these features to drive an unstable point into stable regions, the system stability can be regained.

3.3.4 Learning Algorithm

The learning algorithm is always at the core of an IS. In the literature, various state-of-the-art IS techniques have been adopted, such as Artificial Neural Network (ANN), Decision Tree (DT), Support Vector Machine (SVM) techniques, and sophisticated data-mining techniques, etc. (L. Wehenkel, *et al.*, 1989; L. Wehenkel, *et al.*, 1994; C.A. Jensen, *et al.*, 2001; E.M. Voumvoulakis, *et al.*, 2006; K. Sun, *et al.*, 2007; Y. Xu, *et al.*, 2010; J. Zhao, *et al.*, 2007; I. Kamwa, *et al.*, 2001; I. Kamwa, *et al.*, 2010; I. Kamwa, *et al.*, 2009; N. Amjady and S.F Majedi, 2007; N. Amjady and S. Banihashemi, 2010; A.D. Rajapakse, *et al.*, 2009).

Although there has not been a general principle of choosing a learning algorithm, the accuracy, training speed, robustness, scalability, and interpretability should be considered as problem-specific; in particular, much attention is paid nowadays to *interpretability* for the purpose of obtaining systematic and objective SA rules and control strategies.

3.4 Implementation Stage

While the difficulties in developing the IS can be generally overcome by technically advanced algorithms, to put the IS into practice is even more challenging as it involves not only technical factors, but also the philosophy of the application of IS. The following questions are proposed, which are critical to the success of the real implementation:

- How to make use of the SA results?
- How to reduce misclassification/error?
- How to enhance speed?
- How to improve robustness?
- How to improve the SA&C reliability in practice?

With the above-mentioned merits, ISs have been extensively developed in recent years, and they have broadly shown their effectiveness and efficiency in on-line and real-time stability monitoring. Nevertheless, the existing ISs are still exposed to the following inadequacies in practical stability monitoring applications:

- Due to their statistical learning nature, the ISs inevitably generate uncertain computation errors. Although a large amount of effort has been put to improve their stability monitoring performance, it remains challenging to guarantee the SA reliability.

- In emergency mode, the system can lose its stability within a very short time-frame following the disturbance. Most of the existing ISs for RSA require long post-disturbance observation windows to achieve sufficient SA accuracy, which can otherwise lead to less effective or even ineffective emergency control actions.
- Most of the existing ISs assume perfect (i.e. complete and accurate) data inputs in the intelligent models. However, in practice, the imperfection in measurement data, such as missing-data or measurement noise, can drastically deteriorate the performance of ISs.

3.4.1 Utilize Stability Assessment Results Carefully

Like the situation being encountered in other AI-applied areas, the misclassification/error often exists in spite of best efforts. This can be attributed to the statistical learning essence of AI.

What will happen if operators use an incorrect SA result for operation and control? The consequences can be unnecessary costs, generators and/or load shedding, or even worse, failure to prevent blackouts. The misclassification/error problem can be the most critical factor that hinders the real application of IS to practice today.

It is argued in this research that, in order to be practically applicable, the IS should be able to provide a confident indication of the SA&C result. If the output is sufficiently confident, the operator can adopt the results directly; otherwise, the operator should avoid using such results.

3.4.2 Increase the Accuracy

Continuous efforts are dedicated to increase the accuracy. In addition to using a higher-quality database and significant features as input, there is currently the trend to develop hybrid IS to improve the accuracy, such as (N. Amjady and S.F. Majedi, 2007). In particular, ensemble scheme can be an effective means to increase the accuracy, e.g. *bagging* and *boosting*, which combine a series of k individual classifiers/predictors to complement each other. Since each individual model relies on distinct learning principle, the composite output is considered to be able to reduce the error risk.

3.4.3 Enhance the Speed

In emergency mode, the system can lose its stability within a very short time-frame (normally several seconds) following the disturbance. Most of the existing ISs for post-disturbance SP require fixed and long observation window to achieve sufficient SP accuracy. This means that the SP decisions can only be reliably made with a long delay after the occurrence of the disturbance. Such a delay can result in higher emergency control costs, and lead to less effective emergency control actions. With the above concerns, enhancement in the decision-making speed of IS is crucial in implementing emergency SA&C.

3.4.4 Improve the Robustness

Most of the existing ISs assume perfect input data quality (i.e. complete and accurate measurements, short and consistent communication latency) in the intelligent models. However, in practice, WAMS is exposed to imperfect measurement and communication conditions. Some data quality issues, such as missing-data, measurement noise, and communication delay, can drastically deteriorate the SA&C performance of an IS (V. Terzija, *et al.*, 2011; K. Kirihara, *et al.*, 2014; F. Aminifar, *et al.*, 2012; Y. Wang, 2010; K.E. Martin, 2015). To maintain a satisfactory performance in on-line and real-time SA&C, it is essential to improve the robustness of IS to overcome the practical imperfection in measurement inputs.

3.4.5 Updating of IS for Performance Enhancement

The power system keeps changing in power and topology, which are difficult to accurately forecast in advance. Consequently, there has been a strong need to update the IS to maintain its on-line robustness (see dashed line in Fig. 3.1). Normally, the updating consists of producing new stability information data based on on-line operating information and use of the new data to retrain the IS. At this stage, determination of the timing and the extent to which the model should be updated are important. Besides, the learning speed of the algorithms has been a constraint. It will be meaningless if the training time is too long (say, a few hours to train/tune a neural network).

3.4.6 Integrate with State-of-the-art Tools

The IS can be combined with mature stability analysis tools, such as TDS. In such a condition, the IS could filter the ascertained cases and leave the pending ones to TDS. This way the reliability of SA can be enhanced while the whole computation time can be significantly reduced by filtering a large number of samples.

3.5 Conclusion

This chapter introduced the principles, advantages, and process of applying IS strategy to power system stability analysis. As a promising alternative to conventional methods, IS strategy is advantageous in terms of computation speed, data requirement, interpretability, and knowledge discovery. The critical factors in developing and implementing an IS have been deeply reviewed and discussed. The perspectives on how to overcome the difficulties and challenges in the two stages were discussed, which serve as the baseline for the research.

CHAPTER
4

Intelligent System for On-line Stability Assessment

4.1 Introduction

Modern power systems are currently moving at an accelerated pace towards smart grid era (H. Farahangi, 2010 and E. SmartGrids, 2006), where one of the ambitions is to harness more renewable energies for reducing the environmental impact of the electric power industry (E. SmartGrids, 2006). The increased penetration of renewable power in power systems, on one hand, can considerably reduce CO_2 emissions to counteract the climate changes; on the other hand, the inherent intermittent property of renewables can incur a variety of challenges to the power system. In terms of system operation, unit commitment and dispatching, ancillary services procurement, and frequency control could be affected by the volatile renewable power outputs (Z. Dong, *et al.*, 2010). In the meantime, renewable power can also raise the potential risks to the stability of the power system, especially when the variations of renewable power and load demand coincide (H. Banakar, *et al.*, 2008). One measure to hedge such risk is to perform on-line SA, whose role is to examine the capability of a power system to survive possible contingencies without violating stability criteria. It monitors the security condition of the power system against the risk of blackouts. The risky OPs should be identified accurately, reliably, and well in advance before the preventive control scheme is launched to minimize the risk (P. Kundur, *et al.*, 2004).

Traditionally, SA is performed using TDS, which is too slow to meet the on-line requirement, especially for renewable-dominated power systems. To improve the assessment speed, various ISs based on machine learning techniques, such as ANN, DT, SVM as well as other sophisticated data-mining methods have been applied for power system SA (Z.Y. Dong, *et al.*, 2013; C.A. Jensen, *et al.*, 2001; E.M. Voumvoulakis, *et al.*, 2006; K. Sun, *et al.*, 2007; I. Kamwa, *et al.*, 2001; I. Kamwa, *et al.*, 2010; N. Amjady and S.F Majedi, 2007; N. Amjady and S. Banihashemi, 2010; T.Y. Guo and J.V. Milanovic, 2013; M. He, *et al.*, 2013; H. Sawhney and B. Jeyasurya, 2006;

Y. Xu, *et al.*, 2012; L.S. Moulin, *et al.*, 2004; Y. Zu, *et al.*, 2012; R.S. Diao, *et al.*, 2009; C.X. Liu, *et al.*, 2014; M. He, *et al.*, 2013; K Morison, 2006; Y. Xu, *et al.*, 2011). However, it has also been observed that some state-of-the-art techniques usually suffer from the following issues:

- Most of existing ISs require excessive training time and complex parameters tuning, in particular if a large sized database is learned, making the corresponding SA models difficult to be on-line updated using most recent operating information. As the modern power system is increasingly large, resulting in a growing larger database, there is a pressing demand for more computationally efficient and high-accuracy SA models that can allow more effective real-time implementation and on-line updating.
- Due to the statistical learning nature, the IS strategy for power system SA usually suffers from misclassification or prediction errors. It is widely observed that when the SA output from IS is very close to the classification threshold, the determination of final class could be less accurate. Given this, it is crucial to detect the potential error and avoid the use of such inaccurate SA results. However, the existing ISs lack a reliable mechanism to evaluate the confidence of the results, and thus are not able to distinguish credible and incredible results.
- A tradeoff relationship exists between the SA accuracy and efficiency (i.e. the faster the assessment is achieved, the less accurate the SA results tend to be and vice versa). The existing ISs are not equipped with the ability to interpret such tradeoff or to optimize the SA performance in terms of accuracy and speed. This means the existing ISs are working on sub-optimal conditions to provide SA service, which could increase the system's insecurity risk.
- While most of existing ISs are designed for assessing the transient stability problem, there is yet the IS that focuses on assessment of other power system stability problems, such as voltage stability and frequency stability. With the increasing integration of renewable energy and more active load-side activities, the system dynamics become more complicated and difficult to predict. Therefore, today's power system security should not only be limited to transient stability, but also need to be extended to other stability types, so as to provide comprehensive SA services.

In this chapter, an intelligent SA framework is presented to show how to adopt IS to achieve on-line SA for today's smart grid with considerable wind power penetration. Based on this framework, the above practical issues in IS-based SA are progressively taken into consideration and a series of solutions are provided to deal with them, aiming at integrally optimizing the SA performance of IS and improving the IS adaptivity to various stability problems in smart grid.

4.2 Intelligent Stability Assessment Framework

Benefiting from the huge energy potential of renewable sources and the zero-carbon emission nature, renewable-based generation has been prevailingly implemented around the world. For a power system with large renewable penetration, because of the volatile weather conditions, the system could frequently encounter: (1) fast changes in dispatch of synchronous generators in order to accommodate renewable generation, and (2) fast variations in magnitude and/or direction of power flows through the transmission network. As a result, in order to fully and timely capture the dynamic security conditions in the context of rapidly changing and intermittent operating environments, the SA must be executed based on the current or immediate-future state-of-power systems, and the results should be available on-line within a sufficiently short time frame, say, within a couple of seconds. While the conventional method, such as the TDS is insufficiently fast, an intelligent framework for on-line SA of power systems with large-scale penetration of wind power is developed.

As described in Fig. 4.1, the framework consists of four interactive modules: a wind power and load forecasting (W&LF) engine, a database generation (DBG) engine, a real-time SA engine, and a model updating (MU) engine.

Specifically, the W&LF engine is responsible for predicting the wind power output and the electricity load demand in the system of the

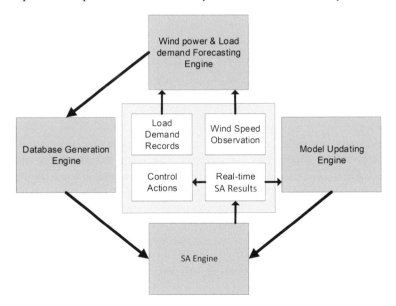

Fig. 4.1: Structure of the intelligent SA framework for wind power penetrated systems

upcoming period over short-term (minutely ahead) and mid-term (daily ahead). With mid-term W&LF results, the DBG engine generates instances, i.e. various possible OPs with corresponding dynamic security index (DSI), subject to credible contingencies. The distributed computing technique can be adopted to accelerate the simulation speed. The SA engine, after being trained by the instances of the DBG engine, is applied on-line for real-time SA. Its input is a set of power system steady-state variables and the output is the DSI with respect to the contingencies. Also, at the on-line stage, W&LF and DBG jointly produce up-to-the-minute instances using the latest operating information, such as load measurements and weather observations. With the latest instances, the MU engine periodically (e.g. hourly) examines the SA performance, and once the SA accuracy is found unsatisfactory, the SA engine will be subject to on-line updating. Performed by the SA engine and at the DBG stage, the SA covers not only the current but also the upcoming time-window of the power system. Benefiting from the generalization capacity and the very fast decision-making speed of the SA engine, the rapid and volatile changes of system's operating state due to the wind power variation can be accommodated comprehensively and on time. Besides, with on-line updating, the system's major changes, including unforeseen wind variations, can be continuously tracked. Based on the SA results, control actions, such as generation rescheduling and load shedding, can be organized to protect the system in case the contingencies really occur. In this framework, the SA engine is the core IS that provides the SA computation, which has been introduced in Chapter 3. The detailed mechanism of SA engine will be presented in the later sections of this chapter. The remaining part of this section only presents the details of the other modules (i.e. W&LF engine, DBG engine, and MU engine) involved in the framework.

4.2.1 Wind Power and Load Demand Forecasting

To be effective, the SA engine should be trained with instances reflecting the possible operating region of the power system. In this regard, it is needed to reliably predict the wind power outputs and the electricity load demand for generating the instances. In the off-line stage, a relatively large number of OPs should be prepared to cover a broad operating region for training the SA engine. Consequently, mid-term (daily ahead) forecasting is conducted. During the on-line SA engine updating stage, short-term (minutely ahead) forecasting is conducted to accommodate the rapid wind-speed fluctuations.

For wind power generation forecasting, the strategy is to predict the wind speed first, and then calculate the wind farm generation by mapping the wind speed against the wind turbine power curve. For this, wind speed forecasting is practically performed in the SA framework.

Generally, wind speed forecasting can be divided into two categories: numerical weather prediction model and data-driven model, where the former models the wind speed within the domain of aerodynamics and the latter relies on the statistic learning of the historical wind speed data. In our research, the data-driven model is employed. For data-driven model, the method can also be classified into two groups: time-series method and AI method. For the time-series models, there are *k-nearest neighbor* (k-NN), *autoregressive integrated moving average* (ARIMA), *generalized autoregressive conditional heteroskedasticity* (GARCH), and kalman filter; for the AI models, ANN, SVM, and RVM methods have been studied in this field(Z. Dong, *et al.*, 2010). The initial results demonstrated that the result from a composite of forecast techniques is often superior to those produced by any individual of the ensemble. The reason is that if the errors in the forecasts produced by different methods are unbiased and have a low degree of correlation with each other, then the random errors from the individual forecast unit will tend to offset each other. Therefore a composite of the forecasts will have lower errors than any individual forecast. As a result, each technique used in the wind speed forecast model will be assigned a weight, which can be adjusted automatically according to its forecast performance.

In the SA framework, a practical load forecasting tool called *OptiLoad* (U. Manual, 2009) and a practical wind speed forecasting tool, both developed at the Hong Kong Polytechnic University, are incorporated for corresponding forecasts. The *OptiLoad* relies on several state-of-the-art forecasting methods, including ANN, SVM, and k-NN for minutely to weekly-ahead load forecasting. During its implementation, the forecasting results provided respectively by the three methods are strategically combined as the final result. According to the practical on-line performance, the weight for each method is dynamically updated. Figure 4.2 shows the graphical user interface of *OptiLoad* (version 1.0b). Its practical daily-ahead forecasting performance on Australia National Electricity Market load data is 1.4~2.1 per cent in terms of the mean absolute percentage error (MAPE).

4.2.2 Database Generation and Model Updating

The database consists of a number of instances, each characterized by a feature vector and corresponding DSI vector. As illustrated in Table 4.1, the feature vector consists of the operating parameters of the power system, such as load, generation output, and bus voltages; the DSI vector describes the dynamic security conditions with respect to different disturbances. Each element corresponds to a contingency, and can be discrete (e.g. 'secure' or 'insecure') or continuous (e.g. security degree).

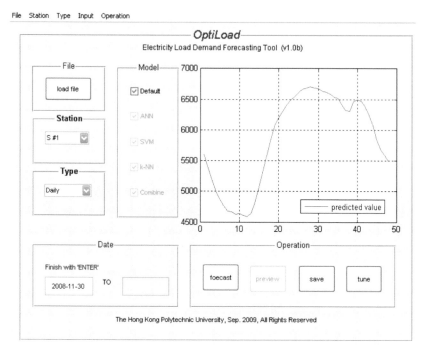

Fig. 4.2: User interface of OptiLoad (v1.0b) (U. Manual, 2009)

Table 4.1: Structure of the stability database

Instance ID	Feature vector				DSI vector			
	F_1	F_2	...	F_n	I_1	I_2	...	I_m
1	-	-	-	-	-	-	-	-
2	-	-	-	-	-	-	-	-
...	-	-	-	-	-	-	-	-
N	-	-	-	-	-	-	-	-

Note: '-' denotes a real value in the database

  The database generation procedure consists of OP production and contingency simulation. With the W&LF results, provided off-line or on-line, together with market transaction information as well as other information where needed, a range of OPs can be produced. To reflect the possible operation region, the forecasted profile should be extended to be made comprehensive. Figure 4.3 illustrates a forecasted profile with its typical forecasting error range; generally the wind speed/load demand within the error ranges should be covered during the OP generation.

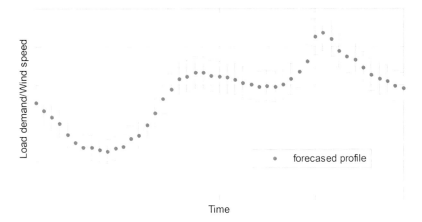

Fig. 4.3: Illustration of W&L forecast

Given the load profile, optimal power flow (OPF) calculation can be carried out to determine the other operational variables, including the active and reactive power outputs of dispatchable generators and voltage magnitudes and angles of buses. Contingency simulation is then conducted on the produced OPs with respect to a pre-defined contingency list, which typically includes the most likely disturbances. Note that the contingency list can also be updated on-line according to the practical operating information. Full TDS can be used to acquire the DSIs of the OPs. It is worth mentioning that there is no essential difference between on-line DBG and off-line DBG. To accelerate the DBG speed, the distributed computation technique is employed in the framework.

For on-line updating of the SA engine, the SA performance can be evaluated by the classification accuracy or prediction errors, known as MAPE or mean absolute error (MAE).

The on-line generated instances are used to continuously (or periodically) examine the SA engine. Once the SA accuracy is found unsatisfactory, it will be subjected to an updating, using the latest instances. This on-line updating process requires the SA engine to have very fast training/tuning speed, so that the updating can be efficiently implemented in the on-line environment to track the variations in data distributions and relationships.

4.3 Database Generation

The application of IS is based on previous knowledge which depicts the characteristics of a system. For power system SA, the stability database should cover sufficient OPs to approximate the practical power system OP space and can withstand uncertain operating condition changes. This section presents a database generation scheme based on extended load profile, extended wind power profile, and economic dispatching, which serves as a methodology that can be adopted by the DBG engine in the intelligent SA framework.

4.3.1 Extended Load Profile

During the operational planning phase, load demand forecasting is conducted, of which the results will be used for unit commitment, OPF, maintenance scheduling, as well as off-line dynamic security assessment, etc. In order to acquire abundant database, the forecasted load profile can be extended for the purpose of enriching OPs.

For a forecasted load profile, the extended active and reactive load demand at each bus is given by

$$P_{Li} = n \cdot P_{Li}^0 \pm \sigma_L \qquad (4.1)$$
$$Q_{Li} = P_{Li} \cdot \eta_L \qquad (4.2)$$

where P_{Li}^0 is the forecast load demand at bus i, n is forecast load profile coefficient describing different load patterns in a range of time intervals, σ_L is the deviation index which is a set of random values representing the unforeseen load pattern changes, and η_L is the reactive power factor. Given a forecasted load profile, the extended load points can be plotted in Fig. 4.4.

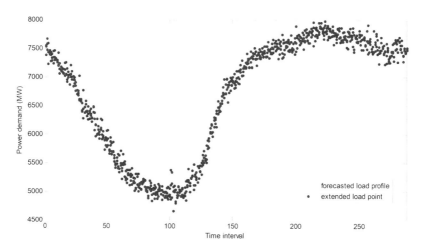

Fig. 4.4: Forecasted and generated load profile

By such a scheme, the extended load points can accommodate a range of changes in load demands during real-time operating phase. And the resulting database tends to be more robust.

4.3.2 Extended Wind Power Profile

Similar to the extended load profile, the extended wind power profile also simulates the uncertainties in the wind power forecast to accommodate the intermittency in the real-time wind-power variations.

For a forecasted wind power profile, the extended wind power profile at each wind farm is generated as follows:

$$P_{Wi} = P_{Wi}^0 \pm \sigma_W \qquad (4.3)$$

$$Q_{Wi} = P_{Wi} \cdot \eta_W \qquad (4.4)$$

where P_{Wi}^0 is the forecast wind power at bus i, σ_w is the deviation index which is a set of variables randomly drawn from the wind-power prediction intervals, presenting the uncertainties in wind-power forecast, and η_w is the power factor of wind-power generation.

4.3.3 Optimal Power Flow and Operating Point Generation

In order to better represent the actual power system operation conditions, other variables of an OP can be determined by OPF scheme, which is usually implemented at the modern dispatching center. The mathematical formulations of OPF model with a generation cost minimization objective can be as follows (J. Zhu, 2015):

$$\min \sum_{i \in S_G} \left(a_i + b_i P_{Gi} + c_i P_{Gi}^2 \right) \qquad (4.5)$$

$$\text{s.t.} \begin{cases} P_{Gi} - P_{Di} - V_i \sum_{j=1}^{n} V_j \left(G_{ij} \cos\theta_{ij} + B_{ij} \sin\theta_{ij} \right) = 0 \\ Q_{Gi} - Q_{Di} - V_i \sum_{j=1}^{n} V_j \left(G_{ij} \sin\theta_{ij} - B_{ij} \cos\theta_{ij} \right) = 0 \end{cases} \qquad (4.6)$$

$$\underline{P}_{Gi} \text{ '' } P_{Gi} \text{ '' } \bar{P}_{Gi} \quad \left(i \in S_G \right) \qquad (4.7)$$

$$\underline{Q}_{Gi} \leq Q_{Gi} \leq \bar{Q}_{Gi} \quad \left(i \in S_R \right) \qquad (4.8)$$

$$\underline{V}_i \leq V_i \leq \bar{V}_i \quad \left(i \in S_B \right) \qquad (4.9)$$

$$\underline{S}_{Li} \leq S_{Li} \leq \bar{S}_{Li} \quad \left(i \in S_L \right) \qquad (4.10)$$

where (4.5) is objective function, (4.6) is the power flow balance constraint set and inequality sets (4.7) to (4.10) are operational constraints; a_i, b_i, c_i are the generation cost coefficients of i-th generator, P_{Gi} is the active output of i-th generator in dispatchable generator sets S_G where the renewable generators are excluded, Q_{Ri} is the reactive output of i-th reactive source in controllable reactive source sets S_R, V_i is the voltage of i-th bus in bus sets

S_B, G_{ij} and B_{ij} are conductance and susceptance between *i*-th and *j*-th bus, respectively, and S_{Li} is the apparent power across *i*-th transmission line in transmission line sets S_L.

Given a set of load points, the above model can be solved to determine the generation output and other variables of an OP. For each generated OP, detailed TDS is executed to identify the DSI with respect to potential contingencies. Each OP can be characterized by a vector of power-system measurable variables, called 'features', and a vector of stability index under contingencies, called 'objects'. In general, the features can be either pre-contingency or post-contingency variables or even their combination, and the object can be denoted by numerical or nominal values.

The database generation scheme can be used both in off-line and on-line. For on-line instance generation, with hourly load forecasting, hourly wind-power forecasting and unit commitment, the prospective OPs for next hour can be generated very rapidly, then distributed computation technique can be adopted to speed up the computing efficiency of detailed TDS. In this way, a large number of on-line instances can be produced.

4.4 Feature Selection

Before training, a data pre-processing stage, known as 'feature selection', is often needed to for the purpose of removing noisy and irrelevant features from the raw database. In doing so, the training data size can be significantly reduced for faster learning speed and better accuracy. The selected significant features can, not only enhance the SA accuracy, but also provide insight into critical system parameters.

In the literature, principal component analysis (PCA) (H. Sawhney and B. Jeyasurya, 2006), divergence analysis (E.M. Voumvoulakis, *et al.*, 2006), Fisher discrimination (C.A. Jensen, *et al.*, 2001), wrapper models (T. wen Wang, *et al.*, 2008), and 'dendograms' evaluation (I. Kamwa, *et al.*, 2009) are amongst frequently adopted methods for feature selection. Although these methods have shown satisfactory performances in eliminating redundant features, some of them are limited to extensive computational burden and/or poor interpretability. For wrapper models, although it can find out a feature subset that results in the best testing performance, the computation burden is always too heavy. PCA is very computationally efficient, but it only gives a linear combination of features from which crucial system parameters are difficult to identify. Fisher discrimination and convergence analysis are based on separability measurements of subsets and tend to provide information on potential system weak points. Therefore, more interpretable and more efficient feature-selection methods are needed to support higher performance for IS-based SA. Two feature selection methods – modified Fisher discrimination criterion and RELIEFF algorithm – are introduced in this section as the candidates for more advanced feature selection. Compared to traditional Fisher discrimination criterion, the modified Fisher discrimination criterion improves its efficiency while retaining its effectiveness. The RELIEFF algorithm is advantageous as it is a transparent and interpretable feature-selection mechanism. These two methods will be also adopted in the rest of the book.

4.4.1 Modified Fisher Discrimination Criterion

As introduced in (C.A. Jensen, *et al.*, 2001), Fisher discrimination measurement is based on Fisher's linear discrimination function $F(w)$, which is a projection from D-dimensional space on to a line in which manner the data is best separated. Given a set of n D-dimensional training samples $x_1, x_2 \dots x_n$ with n_1 samples in class C_1 and n_2 samples in class C_2, the task is to find linear mapping, $y = w^T x$, that maximizes:

$$F(w) = \frac{|m_1 - m_2|^2}{\sigma_1^2 + \sigma_2^2} \tag{4.11}$$

where m_i is the mean of class C_i and σ_i^2 is the variance of C_i. Function $F(w)$ can be written as explicit function of w as:

$$F(w) = \frac{w^T S_B w}{w^T S_W w} \tag{4.12}$$

where S_B is the between-class scatter matrix and S_W is within-class scatter matrix. The class separability of a feature set can be measured by:

$$J_F = \text{trace}(S_W^{-1} S_B) \tag{4.13}$$

The magnitude of J_F can be an index of the linear separability of feature set, the higher the value of J_F, the more separable the data are.

In order to identify the optimal feature subset, the Fisher discrimination measurement is often combined with a searching process (C.A. Jensen, *et al.*, 2006). However, the searching process will introduce a huge computation cost when large-sized data is encountered. This can lead to inefficiency in on-line model updating context. To overcome the drawback in computation speed, Fisher discrimination criterion can be replaced by a simple variation as shown in equation (4.14), which evaluates the discrimination capability of single feature.

For the kth individual feature, its discrimination capability can be calculated by:

$$F_s(k) = \frac{S_B^{(k)}}{S_W^{(k)}} \tag{4.14}$$

where $S_B^{(k)}$ and $S_W^{(k)}$ are the kth diagonal element of S_B and S_w, respectively; the bigger F_s of a single feature, the bigger the discriminating capability it possesses, and the more significance it holds for classification.

For a fast-feature selection, we can calculate F_s for each feature, order them in decreasing sequence of F_s values, and simply select the top ones. As the F_s can indicate the separating ability of a feature in terms of transient stability index, the features with high F_s value are crucial to system stability. The preventive controls can be designed to control these variables to retain system stability.

4.4.2 RELIEFF Algorithm

RELIEFF is a distance-based algorithm that can estimate the importance of features via a transparent and interpretable manner, providing knowledge for other uses, e.g. developing stability control strategies. The idea behind RELIEFF is to evaluate the quality of single features according to how well their values distinguish among instances near each other. It not only considers the difference in features' values and classes, but also the distance between the instances. So the good features can gather similar instances and separate dissimilar ones, providing qualitative estimation on the importance of the features. The original RELIEFF (K. Kira and L.A. Rendell, 1992) consists of iteratively updating the weight for each feature by:

$$W[X]^{i+1} = W[X]^i - diff(X, R_i, H) / N + diff(X, R_i, M) / N \qquad (4.15)$$

where X denotes a feature, R_i is the instance sampled in the i-th iteration, $i \leq N$, H is the nearest instance from the same class as R_i (called *nearest hit*), while M is the nearest instance from the different class with R_i (called *nearest miss*), and N is the number of sampled instance guarantee the weights are in the interval [-1, 1]. Function $diff(X, R, R')$ calculates the difference between the values of feature X for two instances R and R':

$$diff(X, R, R') = \frac{|value(X, R) - value(X, R')|}{\max(X) - \min(X)} \qquad (4.16)$$

From the statistic point of view, the weight of feature X is an approximation of the difference of probabilities:

$$W[X] = P \,(diff.\ \text{value of } X \mid \text{nearest inst. from } diff.\ \text{class})$$

$$- P \,(diff.\ \text{value of } X \mid \text{nearest inst. from same class}) \qquad (4.17)$$

To deal with noises, incomplete data and multi-class problems, RELIEFF can be extended with the following weight updating equation (M. Robnik-Sikonja and I. Kononenko, 2003):

$$W[X]^{i+1} = W[X]^i - \sum_{j=1}^{k} diff(X, R_i, H_j)/(N \cdot k) +$$

$$\sum_{C \neq \text{class}(R_i)} [\frac{P(C)}{1 - P(\text{class}(R_i))} \cdot \sum_{j=1}^{k} diff(X, R_i, M_j(C))]/(N \cdot k) \qquad (4.18)$$

where C is a class label, $P(\cdot)$ is the prior probability of a class, and k is an user-defined parameter.

Instead of finding one nearest *hit* and *miss*, (4.18) finds k nearest *hits* and *misses* to average their contribution in updating the weight, and consequently the risk of miss estimation is reduced. The introduction of $P(\cdot)$ leads to estimating the ability to separate each pair of classes.

For regression problems, i.e. the categorical class is continuous, instead of judging whether two instances belong to the same class or not (absolute difference), the probability of difference is used (M. Robnik-Sikonja and I. Kononenko, 2003). This probability is modeled with the relative distance between the predicted class values of the two instances. Function *diff* (*X*, *R*, *R'*) is reformulated, using the probability that predicts values of two instances:

$$P_{diff\ X} = P\,(diff.\ \text{value of } X \mid \text{nearest instance}) \qquad (4.19)$$

$$P_{diff\ C} = P\,(diff.\ \text{prediction} \mid \text{nearest instance}) \qquad (4.20)$$

$$P_{diff\ C \mid diff\ X} = P\,(diff.\ \text{prediction} \mid diff.\ \text{value of } X$$
$$\text{and nearest instance}) \qquad (4.21)$$

Then $W[X]$ in (4.18) can be re-obtained using the *Bayes* rule:

$$W[X] = \frac{P_{diffC \mid diffX} P_{diffX}}{P_{diffC}} - \frac{(1 - P_{diffC \mid diffX}) P_{diffX}}{1 - P_{diffC}} \qquad (4.22)$$

To identify a set of important features for the IS to further improve accuracy, the features with higher W values can be selected since they can better differentiate the instances among classes. As supplementary benefits, the important features can provide insight into system critical variables as it is valuable for preventive controls.

4.5 Machine Learning Algorithms

In recent years, computational intelligence and machine learning techniques have been widely applied as independent or supplementary tools to facilitate on-line fast SA because of their strong non-linear modeling capability. With these tools, the required computation time to obtain system stability information can be significantly reduced by diminishing the needs for performing TDS. Thanks to advancement in information technologies, many machine learning algorithms have been developed in recent years and are available in the market. In solving the SA problem, although the machine learning tools have demonstrated their exceptional efficiency in solving SA problems, the mechanism of the machine learning algorithms can be very different in nature, which results in deviated performance in their applications. Therefore, the selection of machine learning algorithm is critical in achieving the best SA performance. As mentioned in Section 3.3.4, the accuracy, training speed, robustness, scalability, and interpretability of machine learning algorithms should be highlighted in evaluating their performance in power system SA&C area.

In the literature, ANN, DT and SVM are among the most frequently employed machine learning algorithms for on-line SA. Their mechanisms are briefly introduced in this section and their advantages and disadvantages are also presented. More recently, an emerging family of machine learning algorithm, namely randomized learning algorithms (RLA), has shown much faster learning speed and better generalization performance over conventional algorithms on a number of benchmark problems and engineering applications from regression and classification areas (G.B. Huang, *et al.*, 2006; N.Y. Liang, *et al.*, 2006; G.B. Huang, *et al.*, 2006; G.B. Huang, *et al.*, 2011). Such merits of RLA are also favored in power system SA&C, which also sees its great potential in on-line SA area. Therefore, on top of the conventional machine learning algorithms (i.e. ANN, DT and SVM), two RLAs – extreme learning machine (ELM) and random vector functional link (RVFL) – are also presented in this section. At last, a RLA-based model is developed for on-line SA, and, based on this model, the performance of different algorithms are compared on a benchmark power system.

4.5.1 Artificial Neural Network

The main feature of ANN is its ability of learning from the observed data and encoding the non-linear input-output relationship presented in the data in a retrievable model. ANN can be considered as a network-structured black-box model that is intelligent in the sense that it can map the inputs directly to the outputs, regardless of the complexity of their inherent relationship. The connection of an ANN can be described as a flexible connection of neurons, and each neuron represents a single processing unit that transforms its received inputs into an output through a non-linear process. The structure and the output function of a neuron are respectively shown in Fig. 4.5 and in (4.23), where n is the number of hidden neurons in the network; each w represents a weight value that corresponds to each input signal, b represents the bias value for the neuron, and the activation function works for transforming the linear combination of input into a non-linear relationship function (R.M. Golden, 1997). The commonly adopted activation functions for ANN include Gaussian, sigmoid, hyperbolic tangent (tanh), radial basis function (RBF), rectified linear unit (ReLU), etc.

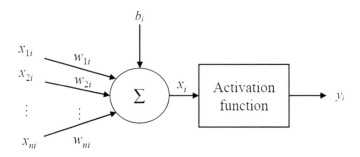

Fig. 4.5: Typical structure of an ANN neuron

$$y_i = f(\sum_{j=1}^{n} w_{ji} x_{ji} + b_i)$$
(4.23)

Due to the flexibility in neuron connections, ANN can be in various forms, such as feed-forward network, probabilistic neural network, adaptive neural network, recurrent neural network, etc. In the literature, the multilayer feed-forward network (MFN), as shown in Fig. 4.6, receives the greatest attention from the power engineering sector. In this structure, the network is represented in multiple layers and each layer is composed of a number of neurons. The outputs of the neurons in each layer are used as the inputs of the next layer neurons. Such feed-forward passage of information continues until the output layer is reached.

The accuracy of an ANN highly depends on the problem to be solved, the available data, and the observed features. Therefore, ANN has to be properly trained with a suitable network structure for good generalization of the entire application. In the typical training process of an ANN, the optimal hyperparameters (i.e. network structure), such as the number of hidden layers, number of neurons in each layer, activation functions, etc. are usually obtained by trial-and-error approach. Generally speaking, a higher number of layers is usually needed for an MFN to solve a more complicated problem or to fit a larger dataset. Under a pre-determined network structure, the parameters involved in the network, including weights and biases, are optimized via back-propagation, which is an iterative process involving three stages:

(1) Feed-forward calculation of the input training patterns with current parameters.

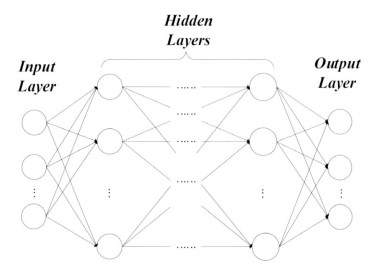

Fig. 4.6: Structure of MFN

(2) Calculation and back-propagation of the fitted mean squared error as follows (R.M. Golden, 1997; S.A.Lorne, 1996)

$$e(\theta) = \frac{1}{2N}\sum_{k=1}^{N}\sum_{m=1}^{M}(y_{k,m} - \hat{y}_{k,m}(\mathbf{x}_{k,m},\theta))^2, \ \theta = \{w,b\} \qquad (4.24)$$

where θ represents the set of weight and bias parameters, $e(\theta)$ represents the fitted mean squared error based on parameter set θ, N is the number of training instances, and M is the number of target outputs for each instance.

(3) Adjustment of the parameters in θ by gradient descent method based on the following equation (R.M. Golden, 1997 and S.A. Lorne, 1996):

$$\theta \leftarrow \left(\theta - \eta \frac{\partial e_t(\theta)}{\partial(\theta)}\right) \qquad (4.25)$$

where η is the learning rate.

The above training process continues until the mean squared error in (4.24) converges to a sufficiently small value. The iterative learning process in back-propagation can proceed in two ways: sequential mode and batch mode. In sequential mode, the parameters are updated after the observation of each training instance, whereas in batch mode, the parameters are updated after the observation of all the training instances in an epoch.

It can be seen that high computation burden is needed for ANN training, at both the hyperparameter setting stage and the network parameter optimization stage. This requires excessive training time and complex parameter tuning, especially when a large-sized stability database is learnt for a large power network. This makes the corresponding SA models difficult to be on-line updated to suit the most recent operating information.

4.5.2 Decision Tree

DT (L. Breiman, 1984), as a popular supervised intelligent learning technique, is a tree-structured predictive model for classification or regression of unknown targets given their features. A DT consists of tree nodes and branches, where branches are the connectors between nodes and nodes can be divided into two types – 'internal node' which represents a feature that characterize an attribute of the problem, and 'terminal node' which denotes a class label of the problem. In structure, each 'internal node' is connected with two successors through an '*if-then*' question while 'terminal node' has no successors.

Figure 4.7 illustrates a typical DT for power system SA. The tree is trained by a stability database, which consists of 1,000 instances each corresponding to an OP (defined by power system operational features) and the stability labels, subject to a contingency (denoted by 'S' class for stable or 'I' class for unstable). The SA of an unknown OP consists of dropping its features (F1 and F2) down the tree from the root node, i.e. Node 1 at the topmost of the tree, until a terminal node is reached along a path. Such a classification procedure can be done very quickly and thus can be applied in an on-line environment to enable earlier detection of the risk of instability.

In developing a DT, the training data is divided into a training dataset and a validation dataset. The growing of the DT starts from identifying an appropriate feature as the root node that can best slip the training instances into distinct classes. Then search begins for all the children nodes again for the best split to separate the remaining subsets. The process is continued recursively until no diversity is increased by splitting the children nodes, yielding a maximal tree. To avoid over-fitting, the maximal tree is then pruned into a series of smaller trees and examined on the training dataset in terms of classification accuracy. The optimal tree is the one giving lowest misclassification cost on the validation dataset. For the tree in Fig. 4.7, there are two internal nodes (where Node 1 is root node) and three terminal nodes.

During the tree induction, the selection of nodes is based on the measurement of the information content (entropy based) that a feature can provide. A feature with higher information content tends to better partition the instances into homogenous regions and thus can be selected as a tree node. Generally, *information gain*, *gain ratio*, and *Gini* are popular

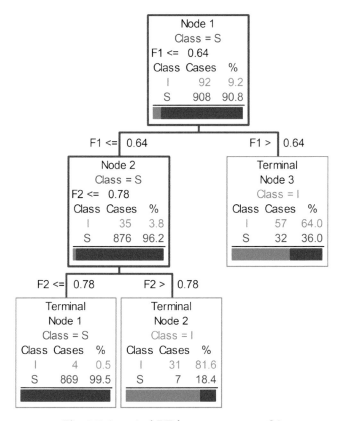

Fig. 4.7: A typical DT for power system SA

algorithms to evaluate the features in DT growing (J. Han, *et al.*, 2011). More technical details regarding DTs could be found in literature (L. Breiman, 1984; J. Han, *et al.*, 2011). The applications of DT on power system SA&C can be found in (L. Wehenkel, *et al.*, 1994; E.M. Voumvoulakis, *et al.*, 2006; K. Sun, *et al.*, 2007; I. Kamwa, *et al.*, 2010; T.Y. Guo and J.V. Milanovic 2013; M. He, *et al.*, 2013; R.S. Diao, *et al.*, 2009; C.X. Liu, *et al.*, 2014; M. He, *et al.*, 2013; E. Karapidakis and N. Hatziargyriou 2002; K. Mei and S.M. Rovnyak 2004; E.M. Voumvoulakis and N.D. Hatziargyriou 2008; I. Genc, *et al.*, 2010).

The attractive advantage of DT over other IS approaches is the ability to provide explicit classification rules, i.e. nodes, their thresholds and associated paths. With these rules, the SA can be transparent and interpretable. Furthermore, the rules provide insight into critical system operating variables relating to stability, i.e. the variables shown at tree nodes. By properly obtaining and interpreting the tree rules, stability control schemes can also be designed for instability prevention. The extracted rules can also be used for post-event auditing and forensic purposes.

4.5.3 Support Vector Machine

SVM was first proposed in 1992 and has attracted a great attention since then (J. Han, *et al.*, 2012). It is mostly applied as a classifier that can map both linear and non-linear data. The working principle of a SVM is to transform the training data into a higher dimension and it searches for the linear optimal separating hyperplane. Such hyperplane serves as the decision boundaries of the instances in different classes, meaning SVM is able to fit complicated non-linear decision boundaries for classification. This feature makes SVM highly accurate on a wide range of classification problems, including SA (L.S. Moulin, *et al.*, 2004; F.R. Gomez, *et al.*, 2011).

When data are linearly inseparable, as shown in Fig. 4.8, it is impossible to use a single straight line to purely separate the classes. In this situation, the original training data needs to be transformed into a higher dimensional space as follows:

$$K(\mathbf{x}_i, \mathbf{x}_j) = \Phi(\mathbf{x}_i) \cdot \Phi(\mathbf{x}_j) \tag{4.26}$$

where $K(X_i, X_j)$ represents the kernel function applied to the original data and $\Phi(X)$ is a non-linear mapping function. The commonly used kernel functions include Polynomial, Sigmoid, and RBF.

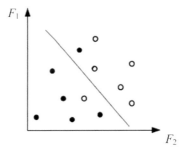

Fig. 4.8: Illustration of a linearly inseparable dataset

The hyperplane is then derived by maximizing the margin between the data instances and the hyperplane, which is equivalent to the following optimization problem:

$$\text{Minimize } \frac{1}{N}\sum_{i=1}^{N}\zeta_i + C\|w\|^2$$

$$\text{where } \zeta_i = \max(0, 1 - y_i(\mathbf{w}\mathbf{x}_i - b)) \tag{4.27}$$

where ζ_i is called the hinge loss function, C is the regularization parameter that controls the tradeoff between maximizing the margin and minimizing the training error. In SVM training, the best kernel function needs to be carefully selected, based on their classification performance on the studied problem and the training data. Moreover, since the values of parameter C and other parameters involved in kernel function can significantly affect the classification performance, they also need to be numerically chosen through grid-search process.

Although SVM has shown great accuracy and generalization ability on a series of classification problems, its computation efficiency is still a concern in real-world applications involving large datasets. The SVM training in high dimensional space generates higher computation burden, and the grid-search process is also time-consuming in finding the optimal model. The above demerits mean that it is technically difficult to perform on-line update of SVM model to fit the most recent data.

4.5.4 Randomized Learning Algorithms

Conventionally, the training protocol of a classical ANN is based on optimizing the parameters, such as weights and biases, to minimize the loss function defined on the desired output of the data and the actual output of the ANN. To be able to map the non-linear input-output relationship existing in most real-world applications, such as SA, the loss function of ANN has to be propagated backwards to guide parameter tuning, which ends up with an iterative learning process. In this situation, such back-propagation training methods are shown to be very time-consuming and may converge to a local minimum. The conventional ANN is also shown to be weak against the ambient noise in realistic applications.

To rectify the above problems, RLA has been developed in recent years. This new family of learning algorithms maps the input-output relationship based on randomly fixed neural network configurations and randomly assigned weights and biases. The advantages of RLA lie in the exceptionally fast learning speed compared to the conventional back-propagation technique and the excellent generalization ability. Two typical RLAs – extreme learning machine (ELM) and random vector functional link (RVFL) – are to be introduced in the sequel. They are the main algorithms employed for the IS presented in this book.

4.5.4.1 Extreme Learning Machine

Concerning on the efficiency of IS in SA application, a new learning algorithm for single hidden layer feed-forward network (SLFN), namely ELM (G.B. Huang, *et al.*, 2006), is considered as a basis to develop more efficient ISs for SA. ELM was proposed by Huang *et al.* as a new learning scheme for SLFNs, which can overcome the insufficiently fast learning speed of conventional learning algorithms. The basic principle of ELM theory is that the parameters of hidden nodes (the input weights and biases for additive hidden nodes or kernel parameters) need not be traditionally tuned by time-consuming algorithms; rather, one can randomly assign the hidden nodes parameters when the activation functions in the hidden layer are infinitely differentiable and analytically determine the output weights (linking the hidden layer to the output layer) through simple generalized inverse operation of hidden layer output matrices.

The structure of a SLFN is shown in Fig. 4.9. For N arbitrary distinct samples $\aleph_N = \left\{ (\mathbf{x}_i, t_i) \mid x_i \in R^n, t_i \in R^m \right\}_{i=1}^{N}$, where $\mathbf{x}_i$ is the $n \times 1$ input feature vector and t_i is a $m \times 1$ target vector. Standard SLFN with $\tilde{N}$ hidden nodes and activation function $\vartheta(x)$ are mathematically modeled as:

$$\sum_{i=1}^{\tilde{N}} \beta_i \vartheta_i \left(\mathbf{w}_i, b_i, \mathbf{x}_j \right) = \mathbf{o}_j, \quad j = 1, \ldots, N \tag{4.28}$$

where $\mathbf{w}_i = [w_{i1}, w_{i2}, \ldots, w_{in}]^{\mathrm{T}}$ is the weight vector connecting the *i*th hidden node and the input nodes, $\beta_i = [\beta_{i1}, \beta_{i2}, \ldots, \beta_{im}]^{\mathrm{T}}$ is the weight vector connecting the *i*th hidden node and the output nodes, and b_i is the bias of the *i*th hidden node. Here $\mathbf{w}_i \cdot \mathbf{x}_j$ denotes the inner product of $\mathbf{w}_i$ and $\mathbf{x}_j$.

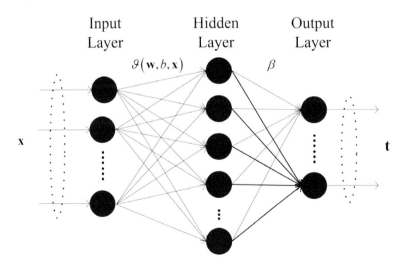

Fig. 4.9: Structure of a SLFN

If the SLFN can approximate these N samples with zero error, we will have $\sum_{j=1}^{N} \left\| o_j - t_j \right\| = 0$, i.e., there exist β_i, b_i and $\mathbf{w}_i$ such that $\sum_{i=1}^{\tilde{N}} \beta_i \vartheta_i \left(\mathbf{w}_i, b_i, \mathbf{x}_j \right) = t_j, j = 1, \ldots, N$ (G.B. Huang, *et al.*, 2006).

The above N equations can be written compactly as:

$$\mathbf{H}\beta = \mathbf{T} \tag{4.29}$$

where,

$$\mathbf{H}(w_1,\ldots,w_{\tilde{N}},b_1,\ldots,b_{\tilde{N}},x_1,\ldots,x_N) = \\ \begin{bmatrix} \vartheta(\mathbf{w}_i,b_i,\mathbf{x}_1) & & \vartheta(\mathbf{w}_{\tilde{N}},b_{\tilde{N}},\mathbf{x}_1) \\ \vdots & \cdots & \vdots \\ \vartheta(\mathbf{w}_i,b_i,\mathbf{x}_N) & & \vartheta(\mathbf{w}_{\tilde{N}},b_{\tilde{N}},\mathbf{x}_N) \end{bmatrix}_{N \times \tilde{N}} \tag{4.30}$$

$$\beta = \begin{bmatrix} \beta_1^T \\ \vdots \\ \beta_{\tilde{N}}^T \end{bmatrix}_{\tilde{N} \times m} \quad \text{and } \mathbf{T} = \begin{bmatrix} t_1^T \\ \vdots \\ t_N^T \end{bmatrix}_{N \times m} \tag{4.31}$$

As named by Huang *et al.* (G.B. Huang, *et al.*, 2006), **H** is called the hidden layer output matrix of the neural network; the *i*th column of **H** is the *i*th hidden node output with respect to inputs x_1, x_2 … x_N. It is proven in (G.B. Huang, *et al.*, 2006) that the input weight vectors $\mathbf{w}_i$ and the hidden biases b_i are in fact not necessarily tuned and the matrix **H** can actually remain unchanged once random values have been assigned to these parameters in the beginning of learning:

(1) If the activation function ϑ is infinitely differentiable, when the number of hidden neurons is equal to the number of distinct training samples, i.e. $\tilde{N} = N$, one can randomly assign the parameters of hidden nodes ($\mathbf{w}_i$ and b_i or kernel parameters). Based on this, analytically calculate the output weights β by simply inverting **H** and therefore SLFNs can approximate $\aleph_N$ with zero error.

(2) While for $\tilde{N} \ll N$ which is the common case in real world, **H** will be a non-square matrix and there may not exist β_i, b_i and $\mathbf{w}_i$. However, one specific set of $\tilde{\beta}_i$, $\tilde{b}_i$, $\tilde{\mathbf{w}}_i$ could be found such that:

$$\left\| \mathbf{H}\left(\tilde{\mathbf{w}}_1,\ldots,\tilde{\mathbf{w}}_{\tilde{N}},\hat{b}_1,\ldots,\hat{b}_{\tilde{N}} \right)\tilde{\beta} - \mathbf{T} \right\| = \\ \min_{\beta_i,b_i,\mathbf{w}_i} \left\| \mathbf{H}(\mathbf{w}_1,\ldots,\mathbf{w}_{\tilde{N}},b_1,\ldots,b_{\tilde{N}})\beta - \mathbf{T} \right\| \tag{4.32}$$

which is equivalent to minimizing the cost function:

$$E = \sum_{j=1}^{N} \left(\sum_{i=1}^{\tilde{N}} \beta_i \vartheta_i \left(\mathbf{w}_i,b_i,\mathbf{x}_j \right) - t_j \right)^2 \tag{4.33}$$

In this case, it is proven in (G.B. Huang, *et al.*, 2006) that, for fixed $\mathbf{w}_i$ and b_i or kernel parameters, equation (4.29) becomes a linear system and the output weights β can be estimated as:

$$\hat{\beta} = \mathbf{H}^\dagger \mathbf{T} \tag{4.34}$$

where $\mathbf{H}^\dagger$ is the *Moore-Penrose generalized inverse* of matrix $\mathbf{H}$. Therefore, one can still randomly assign the parameters $\mathbf{w}_i$ and b_i of hidden nodes and calculate the output weights $\hat{\beta}$ by (4.34) to give a small nonzero training error ε > 0. There are several ways to calculate the *Moore-Penrose generalized inverse* of a matrix, including orthogonal projection method, orthogonalization method, iterative method, and singular value decomposition, etc. (G.B. Huang, *et al.*, 2006).

It can be seen that through such a learning scheme, the time-costly training is not a single calculation step, and as analyzed in (G.B. Huang, *et al.*, 2006), it can still ensure prediction accuracy and may tend to reach better generalization performance. In (G.B. Huang, *et al.*, 2006; N.Y. Liang, *et al.*, 2006; G.B. Huang, *et al.*, 2006; G.B. Huang, *et al.*, 2011), the performance of ELM was comprehensively compared with other state-of-the-art learning algorithms. The results showed that ELM is superior in relative terms. A comprehensive survey of ELM is given in(G.B. Huang, *et al.*, 2011).

4.5.4.2 Random Vector Functional Link

As a RLA, RVFL was proposed in (Y.H. Pao, *et al.*, 1994) and has shown its significance in solving engineering problems (Y. Jia *et al.*, 2014; S. Chai, *et al.*, 2015; Y. Ren, *et al.*, 2016). The structure of a RVFL is illustrated in Fig. 4.10 which consists of an input layer, a hidden layer, and an output layer. Compared to other SLFN, the distinctive feature of RVFL in its structure is the direct links between input and output layers (dashed arrows in Fig. 4.10). With such a structure, both linear and non-linear relationships between inputs and outputs are mapped simultaneously, which statistically improves the prediction performance (Y. Ren, *et al.*, 2016; L. Zhang and P.N. Suganthan, 2016).

Given a database $\{(\mathbf{x}_1, \mathbf{t}_1),\ldots,(\mathbf{x}_i, \mathbf{t}_i),\ldots,(\mathbf{x}_N, \mathbf{t}_N)\}$, where $\mathbf{x}_i = [x_1,\ldots, x_m,\ldots, x_M]^T \in \mathbf{R}^M$ is with M input features and $\mathbf{t}_i = [t_1,\ldots,t_p,\ldots, t_P]^T \in \mathbf{R}^P$ is with P output features, the input-output relationship mapped by a RVFL is (Y. Ren, *et al.*, 2016)

$$t_p = \sum_{j=1}^{J} \beta_{j,p} h_j + \sum_{m=1}^{M} \beta_{m,p} x_m + b_p \tag{4.35}$$

where

$$h_j = f(\sum_{m=1}^{M} \omega_{m,j} x_m + b_j) \tag{4.36}$$

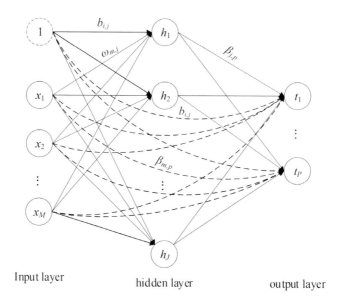

Fig. 4.10: RVFL structure

where J is the number of hidden nodes; f is the activation function; $\omega_{m,j}$ are the weights between input and hidden nodes; $\beta_{j,p}$ are the weights between hidden and output nodes; $\beta_{m,p}$ are the weights between input and output nodes; b_j and b_p are the biases at hidden and output layers, respectively. Equation (4.35) can be rewritten in a compact form:

$$\mathbf{t}_i = \mathbf{h}_i^T \beta, \ i = 1, 2, ..., N \qquad (4.37)$$

where $\mathbf{h}_i$ is the vector concatenating the input features and the hidden nodes of instance i, and β is the output weight vector including $\beta_{j,p}$ and $\beta_{m,p}$.

In RVFL training, the input weights $\omega_{m,j}$ and biases b_j and b_p are randomly assigned within their suitable ranges, so that β can be estimated in a single step (Y. Ren, *et al.*, 2016; L. Zhang and P.N. Suganthan, 2016):

$$\beta = \mathbf{H}^\dagger \mathbf{T} \qquad (4.38)$$

where $\mathbf{H}$ is the matrix of stacking $\mathbf{h}_i$ and $\mathbf{t}_i$ for all N instances in the database, and $\mathbf{H}^\dagger$ is the Moor-Penrose generalized inverse of $\mathbf{H}$, $\mathbf{H}^\dagger = \mathbf{H}^T(\mathbf{H}\mathbf{H}^T)^{-1}$.

It has been proven that RVFL is a universal approximation for continuous functions on finite dimensional sets (L. Zhang and P.N. Suganthan, 2016). Compared to other randomized SLFNs without direct input-output links (G.B. Huang *et al.*, 2006; W.F. Schmidt, *et al.*, 1992), RVFL network achieves better and more stable classification/prediction performance since these direct links serve as a regularization for the randomness in input weight and biases. Another distinctive feature of RVFL network is that the input weights and biases of the hidden nodes are randomly generated. In doing so, conventional back-propagation-based training can be skipped and the training speed of RVFL network can be thousands of times faster. With the above advantages, RVFL classifier is expected to be an effective and efficient intelligent tool for on-line and real-time power system SA.

4.5.5 A Randomized Learning Algorithm-based SA Model

The general framework of IS-based SA models has traditionally been as off-line training and on-line application, wherein the critical part is the learning algorithms. As a matter of fact, many state-of-the-art techniques usually suffer from excessive training time and complex parameter tuning problems in front of large-sized database, which is always encountered in today's bulk power systems.

Also, it is widely known that for supervised learning systems, the more prior knowledge they learn, the better performance they can render. However, the database has been traditionally generated based on the forecasting of prospective operating conditions, while today's power systems are often highly unpredictable in context of restructuring and increasing penetration of renewable energy (especially wind power). This can make the off-line generated stability database insufficient for accurate on-line SA. In (K. Sun, *et al.*, 2007), the authors proposed a periodic checking scheme to partially utilize the on-line acquired information to avoid missing any significant changes of system. But so far, it remains very difficult to make full use of valuable on-line operating information. The major reason can be attributed to the computationally expensive retraining which can be too time-consuming to harvest a practical value.

As presented above, RLA can undertake learning at a very fast speed, based on this an RLA-based SA model is presented in this section. As shown in Fig. 4.11, an initial stability database is off-line generated, based on historical SA archives and off-line instance simulations, and then a feature selection process is performed to identify significant features as input for training. The significant features can not only enhance the accuracy of RLA models but also provide insight on critical system parameters. In the training phase, a set of single RLA models are trained, using the significant features with respect to potential contingencies. This process can be completed very rapidly. During the on-line SA phase, the RLA models take the system snapshot as input and give the SA results within a very short time delay. If unstable status is shown, preventive controls should be activated to retain system stability. In parallel, the current system condition measurements are utilized to generate prospective operating conditions for the upcoming period and full TDS is executed to produce new instances. When the new instances appear unexperienced by the RLA models, they should be used to enrich and/ or update the stability database. In order to obtain more on-line instances, distributed computation technology for TDS can be adopted in this model (K. Meng, *et al.*, 2010). As new instances are generated on-line, the original stability database can be continuously enriched and/or updated, which simultaneously update and/or retrain the RLA models.

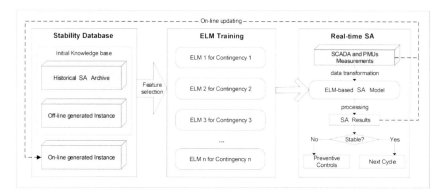

Fig. 4.11: Structure of the RLA-based SA model

It should be noted that the developed model is distinct from traditional computational intelligence-based SA tools for its effective on-line updating capacity, which can enhance the performance in the context of unforeseen major changes in the system. In addition, it is also worth noting that the framework of the presented model is not limited to RLAs, and other machine learning algorithms are also applicable if on-line model updating/enriching is not promptly required. The presented model can also be readily extended to other types of stability problems, provided the problem can be converted into classification or regression form.

4.5.6 Performance Evaluation and Comparison

Based on the presented model, popular machine learning algorithms are compared in terms of computation time and accuracy. The tested algorithms include ELM (a representative of RLA), RBF neural network (RBFNN, a representative of ANN), DT and SVM (K. Sun, *et al.*, 2007; A.D. Rajapakse, *et al.*, 2009; L.S. Moulin, *et al.*, 2004; T. Jain, *et al.*, 2003). These algorithms are respectively tested for SA application which is to classify the unknown OPs into binary classes (stable or unstable) with respect to a contingency. The test system is the New-England 39-bus test system which consists of 10 generators and 39 buses, and its one-line diagram is shown in Fig. 4.12. This is a well known test system for similar stability analysis studies in many previous works.

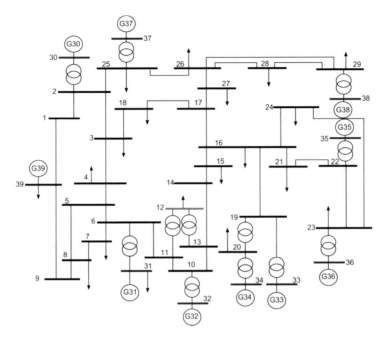

Fig. 4.12: New England 39-bus system

The dataset to test the machine learning algorithms is created as follows: 1,500 OPs are generated using the scheme in Section 4.3, the load level of each bus ranging from 20 per cent to 150 per cent of base load, i.e. $n \in [0.2, 1.5]$, the deviation index σ is given within ±8 per cent, which is consistent with the MAPE of ordinary load forecasting, and the reactive power factor η is given within 0.90 to 0.98. During generating OPs, network topology changes are experienced by randomly removing single or multiple transmission lines in the system. Four severe contingencies are applied including three-phase, two-phase-to-ground, and single-phase-to-ground fault with different clearing time and post-fault topology. The detailed TDS is performed using PSS/E package, and implemented in MATLAB 7.0 running on an ordinary PC with 2.66 GHz CPU. The maximum rotor angle in (1.4) is employed to decide the stability condition in each TDS. The generated database is described in Table 4.2.

In order to examine the proposed method without loss of generality, 500 out of 1,500 generated OPs are randomly sampled with their stability index under the four postulated contingencies as testing data set, and the remaining as training data set.

Test 1: To demonstrate the benefit that on-line updating/enriching stability database can provide, in Test 1, the four algorithms are trained with only 70 per cent of generated training data and evaluated on the entire testing data.

Table 4.2: Generated database

Contingency ID	1	2	3	4
Fault location	Bus 29	Bus 28	Bus 16	Bus 16
Fault type	3-phase	3-phase	1-phase-to-ground	2-phase-to-ground
Duration time	0.14 s	0.14 s	0.18 s	0.16 s
Tripped line	29-26	28-29	16-17	16-21
Stable OP	748	747	610	1248
unstable OP	752	753	890	252

The performance of the four algorithms is shown in Table 4.3. Note that the training time is the overall elapsed time during the training.

Table 4.3: Results of Test 1

Methods	Overall training time (s)	Testing accuracy (%)				
		Fault 1	Fault 2	Fault 3	Fault 4	Overall
ELM	**0.4532**	**96.8**	96.6	94.4	**96.6**	96.1
DT	143.74	96.4	**97.2**	**95.8**	96.0	**96.35**
SVM	1613.4	96.6	96.6	92.2	95.8	95.3
RBFNN	348.98	96.6	96.6	94.4	96.0	95.9

According to the test result, the four algorithms are comparable in terms of testing accuracy (their overall testing accuracy are around 96 per cent, although DT reaches highest). But in computation speed aspect, ELM is far superior over the others: it only consumes less than 0.1 s to learn the training data, while the others cost one to ten minutes.

Test 2: In Test 2, the four algorithms are trained by the whole training data and applied on the same testing data as in Test 1. The performance is given in Table 4.4.

Table 4.4: Results of Test 2

Methods	Overall training time (s)	Testing accuracy (%)				
		Fault 1	Fault 2	Fault 3	Fault 4	Overall
ELM	**0.5001**	**98.6**	**98.8**	97.8	**98.6**	**98.45**
DT	148.62	98.0	98.0	97.8	98.0	97.95
SVM	1727.65	97.2	97.4	96.8	96.6	97
RBFNN	383.64	97.6	97.0	97.0	96.4	97

In the second test, the accuracy of the four algorithms has been improved (overall testing accuracy around 98 per cent, and ELM performs the best in this test). As far as the computation speed is concerned, ELM still shows significant superiority: the training time remains less than 0.1 second, while the other three algorithms cost more minutes. It is anticipated that, for even larger data sets where more contingencies are considered and more features may be encountered, the strength of ELM in computation speed will be more valuable, with which rapid retraining with on-line generated instances can be realized.

For visualized comparison purpose, the overall training time and testing accuracy in the two tests are plotted in Figs. 4.13 and 4.14. Owing to its exceptional SA performance, as shown in the tests, ELM is used as the machine learning algorithms in the rest of this chapter.

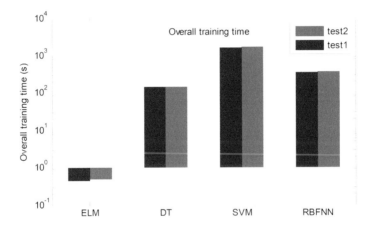

Fig. 4.13: Training time of each method

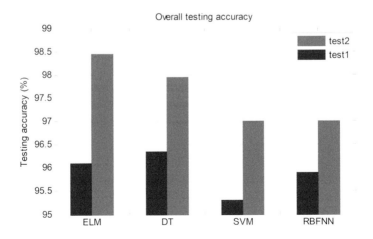

Fig. 4.14: Testing accuracy of each method

4.6 Ensemble Learning

In the literature, a variety of IS techniques have been applied for rapid SA, including ANN, DT, SVM, as well as advanced data-mining approaches. Their effectiveness depends on a proper fitting of the mapping relationship between the power system features and the corresponding DSIs. However, some of them such as ANNs often suffer from excessive training time and/or tedious tuning procedure, which impair their strength for on-line SA. On the other hand, despite best efforts, due to the imperfect fitting and/or insufficient prior information in the training data, most of them can suffer from classification/prediction errors, leading to inaccurate SA results. This cannot be accepted for use, especially in some situations such as under an ongoing disturbance.

To increase accuracy, an effective way is to combine a set of individual learners to make a plurality decision (classification or prediction), known as 'ensemble learning' (L.K. Hansen and P. Salamon, 1990), which will be introduced in this section. Based on the ensemble learning theory, an ELM-based ensemble model is also presented in this section.

4.6.1 Ensemble Learning Theory

Ensemble learning is a learning paradigm where a collection of intelligent learning units are trained for the same classification or regression problem (L.K. Hansen and P. Salamon, 1990; L. Breiman, 2001). In doing so, the single learners can compensate for each other and the whole can reduce aggregated variance and increase the accuracy over the individuals (L.K. Hansen and P. Salamon, 1990; L. Breiman, 2001). In the literature, ensemble learning has been extensively applied to power system stability analysis. For instance, reference (N. Amjady and S.F. Majedi, 2007) reports a kind of ANN ensemble for post-disturbance rotor angle SP, where each ANN individual is trained with post-disturbance rotor angle and velocity values, and the system synchronism is determined via a voting procedure. To improve the efficiency, this model has been modified in (N. Amjady and S. Banihashemi, 2010) with less input features and more effective synchronism status index as output. Also, reference (I. Kamwa, *et al.*, 2010) applies a well-known ensemble-based classifier, called Random Forest (RF) (L. Breiman, 2001). RF combines many single DTs to vote for the final classification decision. Benefiting from both the randomness in tree training and the ensemble effect, RF can be more accurate and stable than single DT. In (I. Kamwa, *et al.*, 2010), the RF is trained using 150 ms and 300 ms WASI features, and significantly increased SA accuracy and reliability are obtained. Besides, by evaluating the class membership possibility of the individual trees, security level can be ranked (I. Kamwa, *et al.*, 2010).

4.6.2 Ensemble Learning for ELM

Although ensemble learning is a general idea, its feasibility and potential benefits fully depend on the property of the individual learning unit. In the meantime, despite the enhanced accuracy, the ensemble could still encounter errors. Keeping this in mind, this section designs an ELM-based ensemble model.

As already introduced, ELM adopts random input weights for learning. Thus it can be absolutely faster than conventional ANNs. As an undesired side effect, the randomness could make single ELM unstable in its implementation stage, but on the other hand, it makes single ELM an ideal candidate for ensemble learning. In our proposal, a series of single ELMs are assembled, each of the individual not only inherently chooses random input weights but also randomizes other parameters in training, including randomly selecting features, training instances, hidden nodes and activation function. Besides the diversity, the specific benefit from the first two randomization schemes is robustness enhancement, and the motivation behind randomizing ELM hidden nodes and activation function stems from the fact that ELM (as well as some other learning techniques) usually has an optimal range of its structural parameters on a given training data. Thus this could improve the accuracy and the generalization capacity.

The ELM-based ensemble learning process can be described as follows:

ELM Ensemble Learning

Given E single ELMs and a database of $F \times D$ size (where F and D are the total number of features and instances, respectively),

For $i = 1$ to E:

(1) Randomly sample d instances out of the database, $1 \le d \le D$.
(2) Randomly pick up f of features in the feature set, $1 \le f \le F$.
(3) Randomly assign this ELM h hidden nodes and an activation function; h is within the optimal range $[h_{min}, h_{max}]$ (subject to a pre-tuning procedure).
(4) Train this ELM and return.

It is necessary to point out that although the designed ELM ensemble is similar to other ensemble-based models, like RF (I. Kamwa, *et al.*, 2010) in structure, it is essentially distinct from them since the ELM, as a kind of unique SLFN, is stochastic in nature. The main advantages of ELM over other algorithms in ensemble learning can be summarized as follows:

- For a successful ensemble model, the key is the learning diversity, including data diversity, structure diversity, and parameter diversity, of the single learning units (Y. Ren, *et al.*, 2016). In constructing the ELM-based ensemble model, data diversity is achieved by randomly extracting subspace from the original database to train individual ELMs. Structural diversity is achieved by randomly assigning the number of hidden nodes of each ELM. Moreover, since the parameters, such as input weights and biases, are randomly assigned in ELM training, parameter diversity is also achieved. Such stochastic nature of ELM makes it advantageous over other learning algorithms in improving the learning diversity of ensemble learning.

- Since multiple learning units need to be trained, the overall training efficiency is also a concern in ensemble learning. In the ELM-based ensemble model, the high training efficiency of ELM can compensate such increased computation burden, which is another advantage of ELM over other learning algorithms in ensemble learning.

- Owing to the high training efficiency of individual ELMs, the ELM-based ensemble model can be on-line updated, using the up-to-the-minute operating information so as to maintain the on-line performance.

- Compared to other algorithms, the generalized randomness in ELM training can fully extend the generalization capability and robustness of the ensemble model.

4.6.3 Performance Evaluation of ELM-based Ensemble Model

A comparative SA test is performed between single ELM units and ELM-based ensemble model to validate the effectiveness of ensemble learning. The test is performed on IEEE 50-machine system (V. Vittal, *et al.*, 1992). This system consists of 50 generators, 145 buses and 453 branches. Classification of transient stability is considered in this test. Following the database generation scheme in Section 4.3, a total of 6,345 OPs of the test system are produced, which cover various load/generation patterns ranging from 20 per cent to 120 per cent of its base OP. The network topologies are changed by randomly removing single or multiple transmission lines at high and/or low voltage levels. A three-phase fault at bus #7 cleared by opening the line between #7 and #6 after 0.1 s is simulated on the generated OPs. Based on the TDS results, 3,916 out of 6,345 instances are secure, and the rest 2,429 instances are insecure.

Based on the original database, RELIEFF algorithm is adopted to select the critical features for ELM training. As a result, from over 500 available features, 50 features are selected to form the important feature set.

Unlike other learning algorithms which have various structural parameters that need to be tediously tuned, the user-defined parameters of ELM-based ensemble model can be simply determined. As the number of single ELM increases, the impact of these parameters can converge to a limit due to the strong robustness of the ensemble model. In this test, we use 200 single ELMs and set the corresponding parameters as follows:

(1) *Sampling parameters*: f and k respectively denote how many features and instances each single ELM randomly samples. In this test, f is set to 30 and k is set to 3,000.

(2) *Hidden neuron node range*: $[h_{min}, h_{max}]$ is the optimal hidden nodes range for single ELMs. It should be determined by a pre-validation procedure. For this, the training data is divided into two non-overlapped subsets, one for training and the other for testing. h_{min} and h_{max} is determined at the highest validation accuracy range. The validation result is illustrated in Fig. 4.15. $[h_{min}, h_{max}]$ can accordingly be set as (K. Meng, *et al.*, 2010); where the single ELM achieves highest accuracy 96.8 per cent.

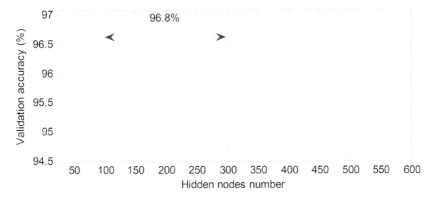

Fig. 4.15: Single ELM hidden nodes vs validation accuracy

(3) *Activation Function*: Sigmoid is used as the activation function of ELM since it results in the highest ELM classification accuracy on the data provided.

The SA performance of a single ELM and an ELM-based ensemble are validated on the provided data through a five-fold cross validation process. Their validated accuracy and computation time are listed in Table 4.5. It can be seen that the accuracy metric values provided by the ELM-based ensemble are all significantly improved over single ELM due to the ensemble effect. Moreover, although more computation time is consumed by ELM-based ensemble, the computational efficiency of the ELM-based ensemble is still very high. With 200 single ELMs on a training data sized 50×5076 (one folder dataset comprises 20 per cent of the total instances, so training data consists of 80 per cent, i.e. 5,076 instances), the total training time of the ensemble model is only 334.5 s, which can efficiently allow on-line updating/retraining. As for the 1,269 testing instances, the ELM-based ensemble costs only 0.68 s to classify all of them, which is faster than even one TDS run.

Table 4.5: Testing results of single ELM and ELM ensemble

Model	Training time (s)	Testing time (s)	Testing accuracy (%)		
			Reliability	Security	Overall
Single ELM	1.54	0.0313	96.21	97.17	96.80
ELM-based ensemble	334.5	0.68	98.85	98.03	98.35

4.7 Credibility Evaluation

It should be noted that for the presented on-line SA model, in spite of the stability database enrichment, a small quantity of misclassification is inevitably encountered. In this section, the misclassified instances are investigated in detail and a credibility evaluation method is introduced to reduce the risk of misclassification during practical implementation.

To demonstrate the distribution of misclassification, the encountered misclassified instances by ELM in the test in Section 4.5.6 are shown in Fig. 4.16 as an example. The vertical axis shows the ELM output values of each misclassified instance.

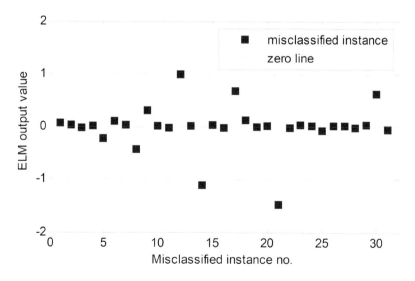

Fig. 4.16: ELM output value of misclassified instance

It can be observed that most of the misclassified instances have an ELM output value very close to 0, which is the threshold value to determine the final stability status. Further investigation on the correctly classified instances shows that their ELM output values have an overall mean value as 0.878 and –0.732 for stable and unstable classifications, respectively. This is an important indication that when the ELM output is very close to the classification threshold, the determination of final class can be less reliable. Based on this finding, the next question is how to accurately classify the instances with unreliable classification results from IS, which is intuitively shown in Fig. 4.17.

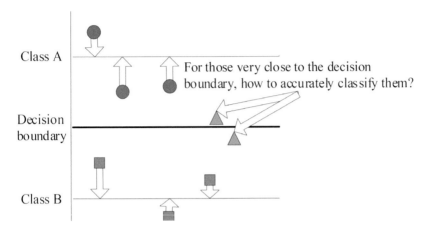

Fig. 4.17: ELM classification mechanism

In order to reduce the risk of misclassification during the practical implementation of the IS, one of the potential methods is to apply TDS to determine an accurate stability status when such 'problematic' output appears. Although in this special case, the computation time can be longer due to the TDS, the probability of appearing such a case is low. So the accuracy can be maintained and the overall computation speed remains very fast. Based on this idea, this section aims to answer another critical research question: how to reliably identify such 'problematic' outputs from IS during an on-line SA process? In this section, the distribution of the correctly-classified and misclassified instances are investigated through an ELM-based SA experiment. Based on the findings from the experiments, a credibility evaluation method is developed and presented to reliably distinguish the potential accurate and inaccurate IS results.

4.7.1 Investigation on Misclassification Phenomenon

In the mechanism shown in Fig. 4.17, the decision boundary serves as the only criterion for classification. In reality, it is probable that some of the instances have ELM output values that are very close to the decision boundary (e.g. the green triangle data points). Even if these instances show marginal results around the decision boundary, ELM still classifies them following the same classification mechanism, which easily leads to misclassification. Therefore, if the marginal instances can be strategically identified and then reasonably classified using more reliable methods, the overall classification accuracy may be improved. Based on the above thinking, a statistical experiment is conducted to further investigate the relationship between the ELM output values and the binary classification accuracy of ELM.

The experiment is performed on a stability database containing 6,257 instances. Each instance consists of the selected features of an OP as its input and the simulated stability status, subject to a contingency as its output (1 represents stable and −1 represents unstable). From the database, 4,000 instances are randomly selected to form the training set while the remaining instances form the testing set.

In this experiment, 200 ELMs are trained as predictors by the training set and applied to classify the OPs in the testing set. The distributions of both correctly-classified and misclassified instances are shown in Fig. 4.18. The horizontal axis represents the regression output range, which is sliced into tiny bins and the vertical axis demonstrates how much percentage of the outputs within each bin are correct (blue) and incorrect (yellow) compared to the simulation results. It can be seen that the probability of misclassification (yellow) is higher when the output is closer to the decision boundary 0, indicating that ELM classification accuracy is positively correlated with the distance between the regression output and 0.

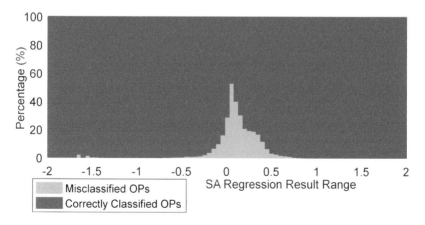

Fig. 4.18: Distribution of accurate and inaccurate ELM outputs

To further verify this finding, another 200 ELMs are trained as predictors by a randomly-generated training set consisting of the same number of instances and the same input features, and either 1 or −1 is randomly assigned as the output in each instance. Such randomly-trained

ELMs are applied to previous testing set. As no knowledge can be derived from a random database, it is rational to regard the prediction outputs of the ELMs as inaccurate. The output distributions of ELMs trained by the power system training set and the random training set are respectively shown in Figs. 4.19(a) and (b). In Fig. 4.19(a), 88 per cent of the ELM outputs are located within [−1.5, −0.5] or [0.5, 1.5], illustrating that trained by a database with relevant knowledge, ELMs tend to generate results close to the predefined class values, −1 and 1. However, in Fig. 4.19(b), 93 per cent of the ELM outputs locating within [−0.5, 0.5] indicate that the prediction results close to 0 tend to be inaccurate, and in another word, incredible. This experimental result coincides with the previous finding in Fig. 4.16. Besides, the regression outputs that are far out of the range of [−1, 1] are also considered incredible because the outliers indicate large prediction errors.

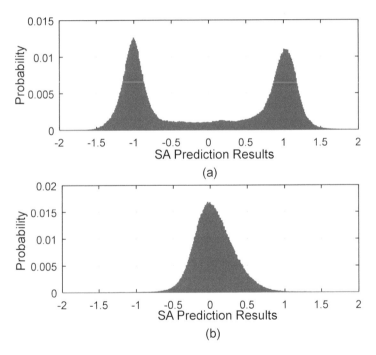

Fig. 4.19: Output distribution of ELMs trained by (a) power system training set, and (b) random training set

4.7.2 Ensemble Decision-making and Credibility Evaluation

In most of the existing ensemble models, the final output, i.e. classification or prediction result, is determined as the voting majority or average value of the individual outputs (N. Amjady and S.F. Majedi, 2007; L.K. Hansen and I. Salamon, 1990; L. Breiman, 2001). These methods are usually not capable of evaluating the credibility of the classification/prediction results. For an ELM-based ensemble model, its advantages lie in not only the accuracy augment but also the possibility to appraise the credibility of the output, with which a reliable and flexible SA mechanism can be allowed. When the output is credible, it can be directly applied as the SA result; otherwise, other schemes, such as TDS, can be resorted for an alternative result. In doing so, the unreliable SA results can be averted for use.

Based on the findings in Section 4.7.1, a new decision-making rule is designed for ELM-based ensemble model in this section to identify the potential error and eliminate the unreliable SA results for use.

Suppose the ensemble model consists of E individual ELMs, we define two decision boundaries for the 'stable' (represented by +1) and 'unstable' (represented by –1) classes as $[lb_s, ub_s]$ and $[lb_u, ub_u]$, respectively. The credible classification rule and credibility estimation of each single ELM is

$$\textbf{If } \begin{cases} y_i \in [lb_s, ub_s] \Rightarrow y_i = 1 \text{ (stable, credible output)} \\ y_i \in [lb_u, ub_u] \Rightarrow y_i = -1 \text{ (unstable, credible output)} \\ y_i \in (ub_u, lb_s) \text{ or } (-\infty, lb_u) \text{ or } (ub_s, +\infty) \Rightarrow y_i = 0 \text{ (incredible output)} \end{cases}$$
$$(4.39)$$

where y_i is the output of the ith ELM unit in the ensemble, $i = 1, 2, \ldots, E$.

A direct estimation of the credibility of the ensemble classification result is the count of its incredible outputs, where, obviously, the larger the count, the less credible it is. Therefore, the ensemble decision-making rule for classification is designed as follows:

Credible Ensemble Classification Rule

Given E single ELMs, which give totally m '0' outputs, s '+1' outputs and u '–1' outputs, $(s + u + m = E)$:

    If $m \geq r \rightarrow$ this classification is incredible

    Else if $\begin{cases} s > u \rightarrow Y = \text{"secure"} \\ s \leq u \rightarrow Y = \text{"insecure"} \end{cases}$

where r ($r \leq E$) is a user-defined threshold to evaluate the credibility of the final classification result Y.

In some other conditions, a continuous value is used to quantitatively measure the dynamic security degree, such as CCT (C.A. Jensen, *et al.*, 2001), stability energy margin (Y. Mansour, *et al.*, 1997), etc., which can be applied in contingency ranking. In this case, there is no prior knowledge about an accurate prediction, but the bias between an individual output and the expectation of the ensemble can be a measure to appraise the potential error of individual prediction. For this, the following credibility estimation criterion is designed for single ELMs.

$$
\textbf{If} \begin{cases} y_i \in [lb_r \times \hat{y}, ub_r \times \hat{y}] \Rightarrow y_i \text{ is credible} \\ y_i \in (-\infty, lb_r \times \hat{y}) \text{ or } (ub_r \times \hat{y}, +\infty) \Rightarrow y_i \text{ is incredible} \end{cases} \tag{4.40}
$$

where $\hat{y}$ is the *median* value of the output vector $[y_1, \ldots y_i, \ldots y_E]$.

It is important to indicate that, in (4.40), the *median* value rather than *mean* value is used to approximate the expected output of the ensemble; the motivation is to keep out outliers that can sometimes be encountered by ELMs. The ensemble decision-making rule for prediction is as follows:

Credible Ensemble Prediction Rule

Given E single ELMs, where there are aggregately m incredible and w credible outputs, respectively ($m + w = E$):

   If $m \geq r \rightarrow$ this prediction is incredible

   Else $Y = \dfrac{1}{w} \sum_{i=1}^{w} y_i$

where r ($r \leq E$) is a user-defined threshold to evaluate the credibility of the final prediction result Y.

It is worth mentioning that in the above rules, for either classification or prediction, only the *credible* instead of *all* the outputs are selected to aggregate as the final result. This, as analyzed in Z.H. Zhom *et al.* (2002), can effectively enhance the accuracy of the ensemble.

4.8 Multi-objective Performance Optimization

In the previous IS that includes the ELM ensemble and the credible decision-making rules, the boundaries to distinguish credible and incredible assessment results are the key parameters that manipulate overall SA accuracy and efficiency. So the next question is how to determine the values of those credible decision parameters (i.e. $\{lb_s, ub_s, lb_u, ub_u, r\}$ for classification and $\{lb_r, ub_r, r\}$ for prediction) to achieve the best classification performance? Such parameters can be empirically selected through trial and error, but this method cannot ensure the optimal performance of the IS. Besides, manual tuning of the parameters is time-consuming and inefficient, especially when the IS is subject to an on-line update. To overcome those inadequacies, this section presents a multi-objective programming (MOP) framework to optimize the SA performance of IS. Under the MOP framework, the IS is able to optimally balance the SA accuracy and earliness. Upon achieving this, the computation time can be further reduced with acceptable impairment in classification accuracy, which enables the early-warning capability of the IS. Therefore, an intelligent early-warning system for on-line SA is developed. This system can also provide multiple options of equally optimal SA performance to the decision makers, who can shift their balanced choice according to the variation of SA requirement.

4.8.1 The Accuracy vs. Earliness Tradeoff

The different parameter values used in ELM ensemble classification can result in different classification performance. Such a phenomenon raises a trade-off relationship between the warning accuracy and earliness in on-line SA application.

In the ELM-ensemble-based IS, the OPs with incredible classification results are re-assessed by TDS which is assumed 100 per cent accurate, so the overall warning accuracy is only determined by the classification accuracy of the credible results. To quantitatively evaluate the performance of reliable ELM ensemble classification, two indices, credibility C and accuracy A, are defined (Y. Xu, *et al.*, 2012):

$$C = \frac{\text{number of credible ensemble results}}{\text{total number of testing instances}} \times 100\% \qquad (4.41)$$

$$A = \frac{\text{number of correctly classified instances}}{\text{number of credible ensemble results}} \times 100\% \qquad (4.42)$$

It can be seen that the number of credible ensemble results acts as both the numerator of C and the denominator of A, so the change of its value affects both the performance indices. Therefore, searching for the best credible decision parameters to optimally separate credible and incredible results is the key to maximize the overall SA classification performance.

The credible decision parameter values should be precisely decided as they determine the performance of reliable ELM ensemble classification. If the credible interval defined by lb_u, ub_u, lb_s, ub_s is too wide and the threshold r is too large, most of the ensemble results will be recognized as credible and the credibility of the ensemble result will be high, but the classification accuracy of the ensemble may not be significantly improved compared to a single ELM. On the other hand, if the credible interval defined by lb_u, ub_u, lb_s, ub_s is too narrow and the threshold r is too small, although the classification accuracy can be extremely high, a large portion of OPs will be assessed as incredible. Since TDS is a time-consuming task, the low credibility of ensemble results implies long computation time, thus a late warning on insecure OPs. Therefore, there is a tradeoff between warning accuracy and warning earliness of SA. The optimization of this tradeoff is an intractable problem.

4.8.2 Multi-objective Programming Framework

As there is a tradeoff between warning accuracy and earliness of SA, achieving optimization on one aspect may not be overall optimal. So a MOP problem is defined as below:

$$\textit{Objectives:} \quad \underset{\mathbf{x}}{\text{Min}} \ \mathbf{q}(\mathbf{x}) = -\mathbf{p}(\mathbf{x}) \tag{4.43}$$

where

$$\mathbf{x} = \ [lb_u, ub_u, lb_s, ub_s, r] \tag{4.44}$$

$$\mathbf{p}(\mathbf{x}) = \ [C, A] = [p_1(\mathbf{x}), p_2(\mathbf{x})] \tag{4.45}$$

where the credibility C and accuracy A are the two objectives which are related to the credible decision parameters through p_1 and p_2 respectively. On top of the objective function, there are also bounding constraints on the elements in $\mathbf{x}$:

$$\textit{Constraints:} \ lb_u < -1, -1 < ub_u < 0, 0 < lb_s < 1, ub_s > 1 \tag{4.46}$$

$$0 < r < 1 \tag{4.47}$$

Although earliness and accuracy are the two final objectives to be optimized in terms of SA performance, credibility C and accuracy A are instead employed as the objectives in the MOP because they are numerical indices and can be mathematically modeled. The optimization of credibility C and accuracy A is equivalent to the optimization of SA earliness and accuracy respectively.

In contrast to a single-objective optimization problem, MOP has two or more objectives simultaneously and its optimal solution is generally not unique if there is a tradeoff between the objectives. Instead, a set of optimized solutions can be provided to the decision maker, who can choose one of them as the optimal solution depending on the practical needs.

The trade-off relationship can be defined by Pareto optimality theory (K. Miettinen, 2012). For the problem (4.43)–(4.47), given a feasible decision space $\mathbf{X}$, a solution $\mathbf{x}^* \in \mathbf{X}$ is called *non-dominated* or *Pareto optimal*, if there does not exist another solution $\mathbf{x} \in \mathbf{X}$ such that $-p_i(\mathbf{x}) \le -p_i(\mathbf{x}^*)$ for at least one of the objective functions, meaning that none of the objective values can be improved without impairment in any other objective. A vector of optimal objective values, $\mathbf{q}(\mathbf{x}^*)$, is called a *Pareto optimal objective vector*. The set of all the Pareto optimal solutions, $\mathbf{x}^*$, is called *Pareto set* and the set consisting of all the Pareto optimal objective vectors is called the *Pareto optimal frontier* (POF).

4.8.3 Intelligent Early-warning System for SA

Combining ELM ensemble classification, credible decision-making rules and MOP, the warning model is developed at off-line stage as shown in Fig. 4.20. As previously mentioned, the critical features are selected from the candidate feature set using proper feature selection techniques, e.g. RELIEFF. Since the proposed MOP problem is built, based on the credible classification results of ELM ensemble, a performance validation process is designed to derive the POF. Eventually, the trained ELM ensemble, the POF and the Pareto set are packaged as the warning model for on-line SA application.

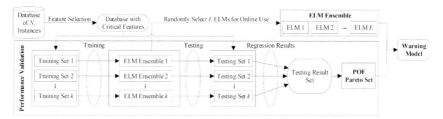

Fig. 4.20: Off-line development of the SA warning model

For the generality of performance validation, k-fold cross validation (J. Han and M. Kamben, 2001) is employed, where the N_i instances in the database are randomly separated into k partitions with an equal number of instances in each partition. Each partition alternately acts as the testing set while the rest $k - 1$ partitions are collected as the corresponding training set. Consequently, the database is rearranged into k training-testing set bundles. Each training set consists of $N_i \times (k - 1) / k$ instances while each testing set consists of N_i / k instances. The k training sets separately train k ELM ensembles. Then the k trained ELM ensembles are applied to their corresponding testing set. In doing so, all the instances in the database are tested. Eventually, there will be $k \times E$ trained ELMs and $E \times N_i$ testing outputs from individual ELMs. All those testing outputs are collected to form a testing result set based on which NSGA-II is employed to solve the MOP problem and search for the POF. Moreover, to alleviate the on-line computation burden, only E ELMs are randomly selected from the $k \times E$ available ELMs to form the ELM ensemble in the warning model.

Based on practical needs and engineering requirements, a compromise Pareto point should be selected from the derived POF. The strategy to make such a selection is: while the accuracy requirement (e.g. ≥ 99.9 per cent) is satisfied, we select the point with the highest credibility. Meanwhile, the compromise credible decision parameters are also decided. It should be clarified that the selection of compromise solution is not restricted to the above strategy. Other decision-making criteria, such as fuzzy logic (S.

Agrawal, *et al.*, 2008) and Nash equilibrium (B. Zhou, *et al.*, 2013) can also be applied where appropriate.

In the off-line development process, the effectiveness of ELM in handling MOP is exhibited. Due to the stochastic nature of ELM algorithm, ELM shows more volatile learning performance compared to other algorithms. In turn, based on the statistical regression results from ELMs, MOP can regulate such volatility to the desired level, and consequently optimize the performance of ELM ensemble. If less volatile learning algorithms are employed in the system, MOP may become less effective in ensemble performance optimization. Additionally, considering the higher training burden of other learning algorithms (recap that ELM generally learns much faster than other algorithms), further MOP can be even infeasible.

How the warning model performs on-line SA is demonstrated in Fig. 4.21. The latest real-time OP is obtained by incorporating the measurement data and the state estimation data from WAMS. The ELM ensemble is then applied to predict the stability status of current OP subject to a contingency. Based on the ELM outputs and the compromise parameters, the final SA decision is made according to the credible classification rule. If an incredible result is decided ($Y = 0$), that specific OP is sent to TDS whose output is used as the final decision instead. If the OP is classified as insecure ($Y = -1$) or TDS delivers an unstable result, a warning of an insecure OP is issued. The whole process continues in real-time to monitor the dynamic security of the system. Moreover, once an incredible result

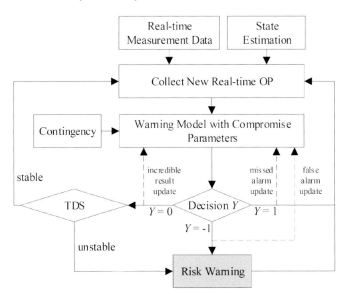

Fig. 4.21: On-line SA application of the warning model

is generated ($Y = 0$) or a false/missed alarm occurs, the warning model is subjected to an update by the latest OP in order to improve its future performance. In the developed early-warning system, each warning model is responsible for one contingency because a different location and severity of the contingency can reform the critical feature set that best reflects the power system security status. Therefore, multiple warning models should be developed off-line for various potential contingencies to be considered. Thanks to the extremely fast learning and tuning speed of ELM, off-line development and on-line update of multiple warning models can be accomplished efficiently. At on-line stage, the multiple warning models are applied recursively to screen all the potential contingencies for each incoming OP. This contingency screening process can be implemented in either a sequential or a parallel architecture.

4.9 Numerical Test on Intelligent Early-warning System

The intelligent early-warning system is tested on New England 10-machine 39-bus system to validate its capability to optimize and regulate the on-line SA performance. Considering the effect of RES on power system dynamics, the power plant on bus 37 is replaced by a wind farm consisting of 376 doubly-fed induction-generator wind turbines. The capacity of the wind farm is 564 MW. The power network used in this test is shown in Fig. 4.22. The power system topology and the dynamic model of the wind turbine are built in PSS/E (M. Seyedi, 2009).

4.9.1 Database Generation and Feature Selection

The stability database is generated based on the variation of wind power and load demand. The power generated by the wind farm on bus 37 ranges from 0 to 564 MW while the demand on each load bus is assumed to vary between 0.5 and 1.4 of their rated values. As a result, the maimum wind penetration in the system is as high as 18.34 per cent and 4,000 OPs are generated and simulated.

TDSs are run to simulate the post-contingency stability status of the generated OPs. The contingencies considered are three-phase faults without loss of any element and they are cleared 0.2 s after the occurrence. Consequently, there are 39 possible contingencies, among which the two listed in Table 4.6 are selected to verify the performance of the developed system. For each contingency, the numbers of secure and insecure OPs are generally not equal. Therefore, to avoid imbalanced learning, resampling technique (A. Estabrooks, *et al.*, 2004) is employed in training individual ELM. The stability status of each OP is represented by the swing margin introduced in Section 1.7.1. Positive and negative TSI respectively indicate

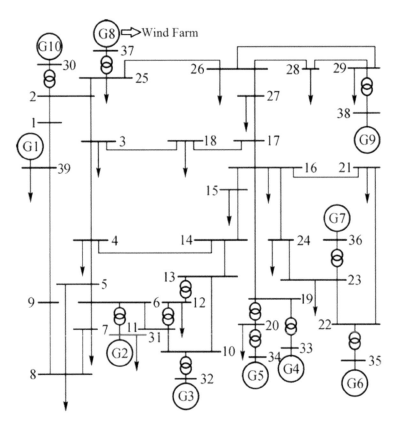

Fig. 4.22: New England 10-machine 39-bus system with wind farm

stable and unstable status. Upon the occurrence of a contingency, an OP with stable TSI is considered secure while an OP resulting in unstable TSI is considered insecure.

Table 4.6: Selected contingencies

ID	Contingency	Stable OPs	Unstable OPs
I	Fault at bus #29	2320	1680
II	Fault at bus #38	1522	2478

By applying RELIEFF algorithm, the candidate features, including the active and reactive output from each generator, load levels, bus voltages magnitudes and angles, line flow, and total generation and load in the system, are ranked based on their importance weights. The feature ranking results for both contingencies are shown in Figs. 4.23(a) and 4.23(b). It can be observed that some features have positive feature

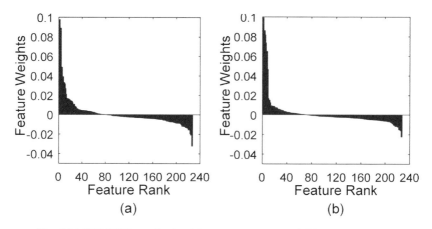

Fig. 4.23: RELIEFF results for (a) contingency I, and (b) contingency II

weights while others have negative values, which respectively mean that they can distinguish or overlap the instances in different classes. Among the 228 candidate features, the top 27 and 42 features with high positive weights are respectively selected as the critical features for contingencies I and II.

4.9.2 Ensemble Learning Parameter Selection

Several parameters have to be tuned and properly chosen in order to have a promising ensemble learning performance.

(1) *Number of ELMs in an ensemble, E*
    In ensemble learning, as the number of neural networks increases, the prediction error decreases but converges to a limit [102, 107]. In the test, E is decided to be 200.
(2) *Optimal hidden node range and activation function*
    The number of hidden nodes and the selection of activation function are the only parameters to be tuned in individual ELM training. Using an activation function, ELM classification accuracy can only reach its maximum within a specific hidden node range. As shown in Fig. 4.24, since the classification performance of different activation functions deviates, only *sigmoid* and *sine* are selected as the activation functions for ELM. [250, 350] are the optimal hidden node range for these two functions.
(3) *Number of training instances, m, and training features, f*
    The decision of how many instances and features are selected to train individual ELMs affects the robustness of the ensemble. In the test, m is chosen to be 2,000 and f is half of the number of critical features.

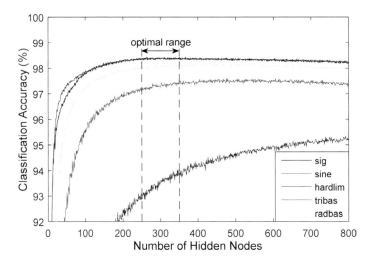

Fig. 4.24: Hidden node and activation function-tuning results

(4) *Number of cross validation fold, k*

The value of k determines the number of ELMs prepared for performance validation. The more the ELMs are available, the better is the robustness. Considering the computational efficiency of performance validation, $k = 5$ is applied.

4.9.3 Performance Validation Result

In performance validation, the MOP problem is solved by NSGA-II and a POF is generated for each contingency. The properties of the POFs are listed in Table 4.7. The worst case credibility (or accuracy) column shows the credibility (or accuracy) while the other performance metric reaches its optima. For contingency I (or contingency II), without any incredible validation result, the accuracy is 98.38 per cent (or 97.92 per cent), while with 100 per cent accuracy, only 89.66 per cent (or 89.64 per cent) of the validation results are classified as credible. POF 1 and POF 2 are plotted in Figs. 4.25(a) and 4.25(b) respectively. It can be seen that the Pareto points form an interpretable and remarkable pattern showing the tradeoff between credibility and accuracy.

Table 4.7: POF characteristics for two contingencies

Contingency	POF ID	No. of Pareto solutions	Worst case credibility	Worst case accuracy
I	1	67	89.66%	98.38%
II	2	94	89.64%	97.92%

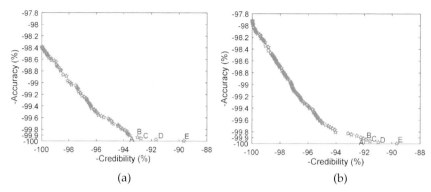

Fig. 4.25: The derived POFs of (a) contingency I, and (b) contingency II

The off-line computational efficiency of the developed system is shown in Table 4.8 where the off-line performance validation is separated into an ELM training process and a NSGA-II optimization process. The ELM training process includes five-fold ELM cross training on a 4000×27 dataset for contingency I and on a 4000×42 dataset for contingency II. The NSGA-II optimization refers to the computation time spent on searching for POF and optimal credible decision parameters. All the computation is performed on a PC with 3.3 GHz CPU. In Table 4.8, for developing an ensemble of 200 ELMs with optimized credible decision parameters, the total computation time is only 374.1 and 424.1 seconds respectively for contingency I and II, demonstrating the high off-line computational efficiency. Moreover, since the developed early-warnings system only needs several minutes for the whole off-line development process, the on-line updating can also be practically performed.

Table 4.8: Off-line computational efficiency (CPU Time)

Contingency	ELM Training	NSGA-II Optimization	Total
I	187.5 s	186.6 s	374.1 s
II	212.5 s	211.6 s	424.1 s

4.9.4 Choose Compromise Parameters

We apply the '3 nines' reliability standard (E.A. Committee, 2008) in this test, meaning a SA accuracy of 99.9 per cent is required. It can be seen that several points in the POFs (labelled as point A to E) satisfy this requirement. A decision has to be made to select the best POF point for on-line application. As the higher credibility can accelerate the contingency screening process and improve warning earliness, among the points with accuracy over 99.9 per cent, the point with the maximum credibility is considered as the best choice. Therefore, the point *A* noted in Fig. 4.25 is chosen as the compromise point for both POF 1 and POF 2. The corresponding credible decision parameters of point *A* are the compromise parameters for the IS.

To be general, the choice of the compromise POF point is not limited to the above strategy. During on-line application, system operators can apply their own decision-making strategies and adjust their choices depending on the SA requirement of the system. Therefore, the developed early-warning system offers the decision makers more flexibility when compared to an IS with fixed parameters.

4.9.5 SA Testing Results

To test the on-line SA performance of the developed IS, another $N_r = 2257$ testing OPs are generated. Similar to the work done in database generation, whether those OPs are secure or insecure is determined by their TSI values delivered from TDS. Such simulated results serve as the actual security status of the testing OPs. By comparing the testing result with the actual security status of each OP, the correctness of each warning activity is examined. In the test, the ELM ensemble, together with the compromise parameters (point *A*), is applied to the testing OPs, and the testing performance and the assessment time for both contingencies are listed in Table 4.9. The testing accuracy for both contingencies is 99.9 per cent, which satisfies the practical SA accuracy requirement. For the 2,257 testing OPs, the warning model takes only 5.12 seconds and 5.04 seconds to complete its task, which shows superior computation speed. Although TDS consumes much longer time, the average processing time spent on each OP is only 0.31 second for contingency I and 0.35 second for contingency II, which is sufficiently short to meet on-line requirement. In Table 4.9, the testing performance using the parameters of the most conservative Pareto point E (100 per cent accuracy) is also demonstrated for comparison. By switching the choice from point *E* to point *A*, the overall computation time is reduced by 44 per cent (from 21.2 min to 11.8 min) and 36 per cent (from 20.7 min to 13.3 min) with an acceptable

Table 4.9: Testing performances on 2,257 testing OPs (CPU time)

Pareto Points	Testing Performance		Computation Time		
	Credibility	Accuracy	Warning Model	TDS	Overall
Contingency I					
A	92.82%	99.9%	5.12 s	11.7 min	11.8 min
E	88.66%	100%		21.1 min	21.2 min
Contingency II					
A	91.61%	99.9%	5.04 s	13.2 min	13.3 min
E	90.19%	100%		20.6 min	20.7 min

accuracy (i.e. meet the 99.9 per cent requirement). This verifies that the warning earliness of the developed system can be significantly improved by optimizing the credible decision parameters and strategically selecting the compromise solution.

4.9.6 Comparison between Validation and Testing Performances

The POF is not only a reference to decide the compromise parameters for on-line use, but also serves as a benchmark for warning model performance. Therefore, by employing any Pareto point on POF, the on-line performance (i.e. credibility and accuracy) of warning model should be close to its validated optimal performance. In Fig. 4.26, the comparison between the testing performance at each available Pareto point and the validated POF for contingency I is demonstrated as an example. It can be seen that the testing performance points are distributed around POF and follow the pattern of POF.

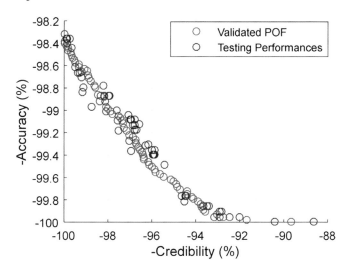

Fig. 4.26: Validation and testing performances for contingency I

Besides, mean radial error (MRE) is employed to evaluate the error between POF and the testing performances (G.R. Hancock, *et al.*, 1995):

$$RE = \sqrt{(C - \hat{C})^2 + (A - \hat{A})^2}$$

(4.48)

$$MRE = \overline{RE} = \frac{1}{N_r} \sum_{i=1}^{N_r} RE_i$$

(4.49)

where *RE* stands for radial error which is the geometrical distance between testing and validation performance; *C* and *A* are the credibility and accuracy of the testing results while $\hat{C}$ and $\hat{A}$ are for Pareto points.

Moreover, MAE is used to demonstrate the error on each single index:

$$MAE_C = \frac{1}{N_r} \sum_{i=1}^{N_r} (C - \hat{C}) \tag{4.50}$$

$$MAE_A = \frac{1}{N_r} \sum_{i=1}^{N_r} (A - \hat{A}) \tag{4.51}$$

The *MRE*, MAE_C and MAE_A between the testing and validation performance for both contingencies are listed in Table 4.10. The maximum *MRE* is as small as 0.41 per cent and 0.39 per cent for contingency I and II respectively. The MAE_C and MAE_A are also much less than 0.1 per cent for both contingencies. Overall, the error between validated performance and on-line performance is considered insignificant and acceptable.

Table 4.10: Error between validation and testing performances

Contingency	*MRE*	MAE_C	MAE_A
I	0.41%	0.024%	0.069%
II	0.39%	0.011%	0.072%

4.9.7 On-line Updating Verification

To verify the on-line updating capability of the IS, a test is conducted to simulate the updating process. Besides the full database, performance validation is also performed based on 75 per cent and 50 per cent of the instances in the database. The derived POFs for contingency I are shown in Fig. 4.27. It can be seen that the obtained POFs are moving front with more training instances, which clearly verifies that the updating of the system definitely improves the warning accuracy and earliness. As a warning model is for one contingency, the consumed updating time for each model should be comparable to the off-line computation time shown in Table 4.8. It is worth noting that, according to the validated credibility of the selected POF point *A*, only less than 10 per cent of the OPs are incredibly assessed, in turn indicating that only a small portion of the models are subject to update for a specific OP. Moreover, as the on-line updating process continues, the credibility of the system should be improved. Thus even fewer models will need to be updated and the updating process will be even faster. Therefore, considering that distributed computation architecture (K. Meng, *et al.*, 2010) can be used to further improve the computational efficiency, the updating process is fully compatible with on-line SA requirement.

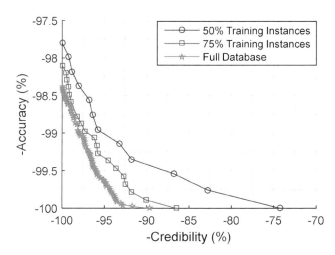

Fig. 4.27: Model updating verification for contingency I

4.10   Hierarchical IS for On-line STVS Assessment

As one essential security criterion, dynamic security refers to the ability of a power system to survive the dynamic transition from an initial operating equilibrium to a viable new equilibrium with most limits bounded after being subject to a disturbance (P. Kundur, *et al.*, 2004). To protect a power system against the risk of blackouts, SA needs to be performed to examine various stability criteria, including transient stability, voltage stability, and frequency stability.

While the IS presented in Section 4.8.3 is only designed for transient stability problem, this section focuses on another yet increasingly important dynamic security concern – STVS, and develops a hierarchical IS for its on-line assessment.

Unlike rotor angle stability, STVS involves fast and complex load dynamics, which is even more difficult to capture. The instability can propagate very rapidly and trigger significant load shedding or even wide-spread blackouts. Nowadays, due to the increased penetration of wind power, the STVS is becoming a critical concern in terms of low voltage ride through (LVRT) capability for wind turbines. To address the STVS assessment problem, we employ ELM and ensemble learning strategy to design a hierarchical IS. The STVS assessment problem is modeled as a *transient voltage collapse* sub-problem, which corresponds to a classification process and an *unacceptable dynamic voltage deviation* sub-problem, which corresponds to a regression process. The two sub-problems are sequentially handled by two ensemble-based classifier and predictor models.

4.10.1 Short-term Voltage Instability: Mechanism and Phenomenon

The IEEE and CIGRE joint task force defines the voltage stability as the ability of a power system to maintain steady voltages at all buses in the system after being subjected to a disturbance from a given initial operating condition (P. Kundur, *et al.*, 2004). As a major threat to power system security, the loss of voltage stability can result in progressive voltage drop or rapid voltage collapse, which can trigger significant load shedding or even wide-spread blackouts. Over the past decades, many major blackouts around the world have been found directly associated with this phenomenon (T.V. Custem and C. Vournas, 1998; C.W. Taylor, *et al.*, 1994; J.D. De Leon and C.W Taylor, 2002).

In general, voltage instability stems from the attempt of load dynamics to restore the power consumption beyond the capability of the combined transmission and generation system (T.V. Custem and C. Vournas, 1998). According to (P. Kundur, *et al.*, 2004), the voltage stability can be divided into long-term and short-term phenomena. The long-term voltage stability involves slower acting equipment, such as tap-changing transformers, thermostatically controlled loads, and generator current limiters (P. Kundur, *et al.*, 2004). It can be modeled as algebraic equations and solved by power flow-based methods (T.V. Custem and C. Vournas, 1998). Therefore, its computation speed can be fast enough to enable (near) real-time implementation. By contrast, the STVS phenomenon involves complex dynamics of load components, such as induction motors tending to restore their consumed power in a very short time-frame (say, several seconds) (J.D. De Leon and C.W Taylor, 2002). Following a large disturbance, the induction motors decelerate dramatically by the voltage dip or may stall if the electrical torque cannot overcome the mechanical load. This in turn draws very high reactive current which affects adversely the voltage magnitudes. As a consequence, unacceptable transient voltage performance (e.g. FIDVR) may be experienced, and/or transient voltage collapse may occur. With increased penetration of induction motor loads (e.g. air-conditioners), today's power systems tend to be more vulnerable to short-term voltage instability (J.D. De Leon and C.W. Taylor, 2002).

4.10.2 Dynamic Load Modeling

As mentioned above, the key driving force of voltage instability is the tendency of dynamic loads to restore their consumed power within a very short time frame. Therefore, modeling dynamic loads is a necessity for STVS study.

A composite load model 'CLOD' has been defined in the commercial power system simulation package PSS®E to represent various dynamic load components, which can be used in STVS study. As illustrated in Fig.

4.28, this load model is an aggregation of large motors (LM), small motors (SM), discharge lighting, transformer exciting current, and constant power load, as well as the voltage-dependent load, all of which are fed from many real-world substations. However, other dynamic load models can also be used if necessary.

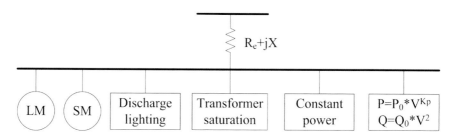

Fig. 4.28: Structure of the composite load model 'CLOD'

The mathematical model of induction motors is also represented by DAEs. Hence, the whole mathematical model of the power system for STVS study composes a very large set of non-linear DAEs which can be compactly formulated as follows:

$$\frac{d\mathbf{x}}{dt} = \mathbf{f}(\mathbf{x}, \mathbf{y}, \mathbf{p}, \lambda) \tag{4.52}$$

$$0 = \mathbf{g}(\mathbf{x}, \mathbf{y}, \mathbf{p}, \lambda) \tag{4.53}$$

where set (4.52) corresponds to the differential equations of the system components, including generators, motors, dynamic loads, as well as their control systems, etc.; set (4.53) corresponds to the algebraic equations of the network and static loads. Vector $\mathbf{x}$ denotes the state variable (e.g. generator angles and dynamic bus voltages); vector $\mathbf{y}$ denotes algebraic variables (e.g. static load voltages and angles); vector $\mathbf{p}$ stands for controllable parameters (e.g. automatic voltage regulator (AVR) set-points); and vector λ represents uncontrollable parameters (e.g. load levels).

The DAEs cannot be analytically solved. Rather, one can solve it via step-by-step integrations. In practice, the STVS assessment is realized based on TDS, which is, however, very computationally burdensome.

4.10.3 Simulation Study on Dynamic Load Impact

Using the composite load model 'CLOD', we examine the impact of load dynamics on the short-term voltage performance of a power system. Figs. 4.29 and 4.30 illustrate the dynamic voltage response in the New England 39-bus system following short-circuit disturbances with different severity degrees (Y. Xu, *et al.*, 2014).

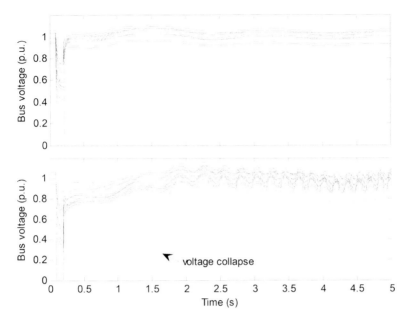

Fig. 4.29: Dynamic voltage responses without (upper window) and with (lower window) load dynamics – a very severe fault

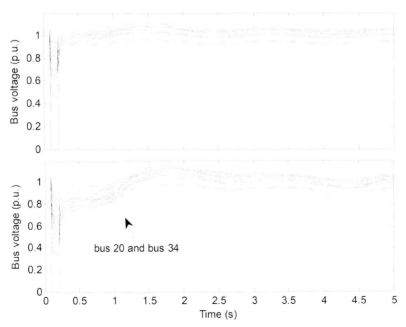

Fig. 4.30: Dynamic voltage responses without (upper window) and with (lower window) load dynamics – a less severe fault

The voltage trajectories are obtained via TDS (Y. Xu, *et al.*, 2014). For each figure, the upper and lower windows respectively show the voltage response without the load dynamics (i.e. the dynamic load model is not used, and a constant power load is used), and with load dynamics (i.e. the dynamic load model is used). For each figure, it can be seen that without load dynamics, the bus voltages recover very fast to the normal level and the system is stable. In Fig. 4.29, a very severe disturbance is applied, and the lower window shows that transient voltage collapse occurs at around 1.6 s. The voltage collapse is an unacceptable situation for a practical power system, since it can lead to more serious consequences, such as cascading failures and/or blackouts. For Fig. 4.30, a less severe disturbance is applied, and the lower window shows that some buses (in particular bus 20 and bus 34) experience a prolonged voltage depression and delayed voltage recovery. In practice, the delayed voltage recovery can trigger undervoltage load shedding (UVLS) devices that curtail a significant amount of loads.

4.10.4 The Two Sub-problems in STVS Assessment

As illustrated in Figs. 4.29 and 4.30, STVS study mainly focuses on (1) fast voltage collapse – *see* Fig. 4.29 or (2) unacceptable post-disturbance voltage deviation – *see* Fig. 4.30. Therefore, the STVS assessment problem can be modeled as a transient voltage collapse sub-problem and an unacceptable dynamic voltage deviation sub-problem. The first sub-problem checks the existence of a short-term equilibrium of the post-contingency system and the second sub-problem quantifies the degree of stability.

To evaluate the short-term voltage performance, proper stability criteria should be used for each sub-problem. The evaluation of transient voltage collapse is a binary question, that is, whether or not the system loses the short-term equilibrium following a disturbance. Consequently, the TVCI presented in Section 1.7.2.2 is used to signal the occurrence of transient voltage collapse. After the system withholds a short-term equilibrium, the system may still experience unacceptable dynamic voltage deviation. Actions of motor protective devices can trip a significant amount of loads. In contrast to transient voltage collapse, the severity of unacceptable dynamic voltage deviation can vary depending on the severity of a disturbance. In this case, a continuous index must be applied. Therefore, the TVSI presented in Section 1.7.2.2 is employed to quantify the unacceptable dynamic voltage deviation sub-problem.

Both the indices can be calculated via performing TDS to obtain the voltage trajectories, while the TVCI is a binary value and the TVSI is a continuous value. Taking the simulation results in Section 4.10.3 as an example, the TVCI is 0 for Fig. 4.29 (lower window), and the TVSI is 0.37 for Fig. 4.30 (lower window).

4.10.5 Hierarchical Intelligent System

The on-line STVS assessment can be converted as a classification sub-problem for transient voltage collapse and a regression sub-problem for unacceptable dynamic voltage deviation. The latter depends on the result of the former. Therefore, a hierarchical IS is developed for on-line STVS assessment.

The hierarchical IS is based on ELM ensemble model with different output aggregation strategies used for the two sub-problems. Average value of individual ELM outputs is used for the regression problem (i.e. transient voltage collapse sub-problem), whereas the majority class of the individual ELM outputs is used for the classification problem (i.e. unacceptable dynamic voltage deviation sub-problem).

The STVS assessment routine of the hierarchical IS is as follows: in the off-line stage, a voltage stability database is prepared by performing comprehensive TDS considering various OPs and/or collecting disturbance records. Then, the IS is trained by the database. In the on-line application stage, once the input is available, the STVS can be assessed by the IS rapidly and the results are used to support control decision-makings. Also in the on-line stage, with updated system information (e.g.network structure, load composition, etc.), a new stability database can be generated to update/enrich the IS. Note that this on-line updating aims to keep the IS adaptive to unexpected system changes.

The hierarchical IS for on-line STVS assessment is conceptually described in Fig. 4.31. The IS consists of an ensemble classifier and an ensemble predictor based on the ELM ensemble model. They are pre-trained in off-line stage with forecasted OPs (e.g. day ahead) and corresponding voltage collapse status and TVSI. For on-line application, the real-time system measurements (including generator outputs, bus loads, bus voltage magnitudes, bus angles, etc.) are collected by SCADA or PMU devices. Once the input measurements are obtained, they are firstly dropped to the classifier, which determines the transient voltage collapse condition of an unknown OP with respect to a certain disturbance. If voltage collapse is ascertained, remedial control actions should be taken to protect the power system from the voltage collapse should the disturbance really occur; otherwise, the measurements are then dropped to the predictor to evaluate the TVSI of the OP. If the predicted TVSI is unacceptable, e.g. less than a certain threshold, remedial actions will be taken. During the on-line application phase, the classifier and predictor can also be updated using the latest on-line operating information.

Due to the very fast training speed of ELM, the IS can be updated on-line. The updating can be done periodically, e.g. hourly or half-daily, depending on practical needs. The data for updating can be obtained from TDS results on the real-time system OPs. The simulation results can be

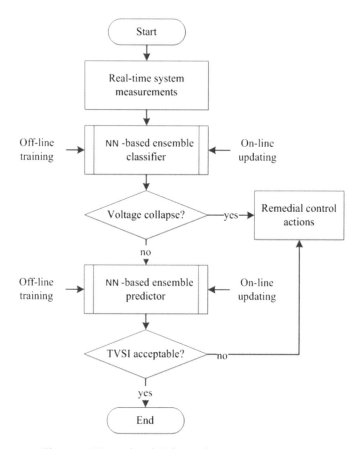

Fig. 4.31: Hierarchical IS for on-line STVS assessment

saved as new training data. Such new training data can be used as a new set or attached to the initial training database to retrain the IS.

4.11   Numerical Test on Hierarchical IS

The developed hierarchical IS is tested on the New England 10-machine 39-bus system (*see* Fig. 4.22) which is a popular test system for stability study (Y. Xu, *et al.*, 2012).

The numerical simulation is conducted on a 64-bit PC with 3.1 GHz CPU and 4.0 GB RAM. TDS is performed using commercial software PSS®E (Y. Xu, *et al.*, 2014), and the developed hierarchical IS is realized in the MATLAB platform. For the test system, the dynamic load parameter is assumed as: LM – 20 per cent, SM – 30 per cent, discharge lighting – 5 per cent, transformer exciting current – 5 per cent, constant power load – 30 per cent, voltage-dependent load – 10 per cent.

4.11.1 Database Generation

To train the hierarchical IS, a comprehensive stability database is necessary. In practice, this can be obtained by performing simulations on various scenarios and/or fault recordings. In this study, we artificially generate such a database through OPF calculation and TDS (R. Zhang, *et al.*, 2011). A total of 700 different OPs covering a variety of different load/generation patterns ranging from 80 per cent to 120 per cent of the base loading level are generated through OPF calculation. A three-phase short-circuit fault is considered first (denoted as fault #1). The fault is applied to bus 15 with a duration time of 0.2 s. For each OP, the following candidate features are selected as the input vector: P, Q load, voltage magnitude and angle of each load bus; P, Q generation of each generator; total P, Q load and generation of the whole system. For each OP, there are a total of 136 candidate features.

TDS is then performed on each generated OP to determine their TVCI and TVSI. The total simulation time is 5s and the simulation step is 0.01s. In calculating the TVSI, μ is set to 20 per cent. For the 700 generated instances, 256 of them (36.6 per cent) experience voltage collapse and 444 (63.4 per cent) do not experience voltage collapse.

The structure of the STVS database is illustrated in Table 4.11. Note that the input features are normalized into the range of [0, 1]. For TVCI, 1 represents the voltage collapse occurs while 0 not. Only the samples with 0 TVCI have a TVSI. The histogram plot of the TVSI of the samples is given in Fig. 4.32, where it can be seen that the range [0.3, 1.5] occupies the major proportion.

Table 4.11: Structure of the STVS database

Sample ID	Input vector				Output vector	
	F_1	F_2	...	F_n	TVCI	TVSI
1	0.0	0.0	0.0	0.0	1	n/a
2	0.22	0.13	0.22	0.64	0	2.6
...	...	...	...	...	...	...
N	1.0	1.0	1.0	1.0	0	3.8

Note that during the training phase the classifier is trained by all of the instances and the TVCI is used as the training target, while the predictor is only trained by the instances whose TVCI is 0 and only the TVSI is used as the training target.

Given the 700 instances, we randomly pick up 25 per cent, i.e. 175 instances as the testing set, leaving the remaining 525 instances serve as training set.

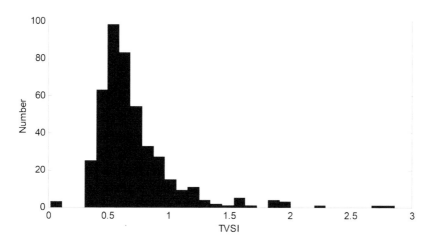

Fig. 4.32: Histogram plot of the TVSI of non-voltage collapse instances

4.11.2 Feature Selection

The RELIEFF algorithm is used to select the significant features. The RELIEFF values of the 136 features are shown in Fig. 4.33, where it can be seen that different features have varying capabilities to distinguish the OPs in terms of the status of STVS. A large RELIEFF value indicates a stronger capability. The top three features are the voltage angle of bus 24, Q load of bus 27, and voltage angle of bus 26. All of them are voltage related variables of the system. In the following test, we use the top 20 features as the model inputs.

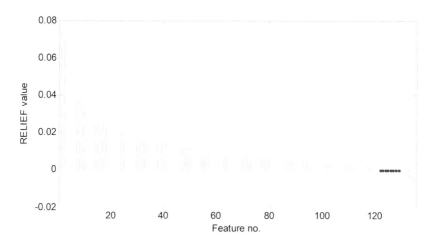

Fig. 4.33: RELIEFF value for each candidate feature

4.11.3 Model Tuning

According to the ELM theory, a certain hidden node number should be predetermined before training. This corresponds to the optimal hidden node range for the ELM ensemble. Since the hidden node has a direct impact on the accuracy of a single ELM, the optimal hidden node range needs to be well tuned in advance. For this purpose, the training data set is randomly divided into two non-overlapped sets – one for training and the other for validation. The optimal hidden node range can be determined at the range where the lowest validation error is obtained. To measure the performance of the hierarchical IS, percentage classification accuracy is used for the classifier, and the MAPE is used for the predictor:

$$MAPE = \frac{\sum_{i=1}^{d}\left|y_i - \hat{y}_i\right|}{\left|y_i\right| \cdot d} \tag{4.54}$$

where y_i and $\hat{y}_i$ are the actual and predicted TVSI value, respectively, and d is the total number of the testing instances.

The tuning profiles are shown in Figs. 4.34 and 4.35. It can be seen that for the classifier, with continuously increased hidden nodes, the classification accuracy increases sharply but then reduces gradually. The highest classification accuracy range occurs when 30 to 60 hidden nodes are used. For the predictor, a similar trend is also observed (with continuously added hidden nodes, the prediction error decreases sharply and then increases gradually), and when 15 to 50 hidden nodes are used, the resulting MAPE is at the lowest range. Therefore, these hidden node numbers are used to train the ensemble model.

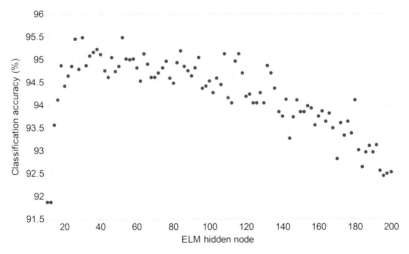

Fig. 4.34: Tuning profile for the classifier

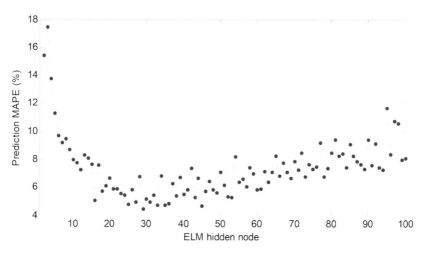

Fig. 4.35: Tuning profile for the predictor

Also from Figs. 4.34 and 4.35, it can be observed from the tuning process that a single ELM can achieve maximum 95.5 per cent classification accuracy and minimum 4.56 per cent prediction MAPE. Besides, it is worth mentioning that the training of a single ELM is very fast, it only costs 0.01 seconds to learn the tuning data set.

4.11.4 Performance Test

In this test, we use 100 individual ELMs to build the ensemble model. Once the model is well trained, it can be applied on-line. As soon as the input variables (significant features) are available (e.g. measured by PMUs), the STVS can be determined instantaneously.

The testing set is used to validate the hierarchical IS. For comparison purpose, the single ELM is also tested. Besides, some state-of-the-art single learning algorithms, including RBFNN, SVM and DT, are also tested with the same training and testing data sets. The performance of the algorithms is presented in Tables 4.12 and 4.13, respectively.

Table 4.12: Performance in terms of accuracy (Fault #1)

Model	Classification	Prediction MAPE
Hierarchical IS	98.6%	1.82%
Single ELM	95.3%	3.83%
SVM	95.8%	3.96%
DT	96.2%	4.68%
RBFNN	94.1%	2.96%

Table 4.13: Performance in training efficiency (Fault #1)

Model	Training time (s)
Hierarchical IS	28
SVM	78
DT	16
RBFNN	62

As Table 4.12 shows, the developed hierarchical IS outperforms the others in both classification and prediction performance. Importantly, it should be noted that the hierarchical IS significantly increases the performance over the single ELM, achieving very high classification accuracy and very low prediction error.

As Table 4.13 shows, the training efficiency of the hierarchical IS is very high, although 100 ELMs are used. The overall training time is 28 s, depending on the number of hidden nodes, one single ELM only cost around 0.3s for training. This is due to the unique learning process of the ELM – randomly selecting the input weights and analytically determine the output weights. Compared with other learning algorithms, DT is fastest since its learning procedure does not require time-consuming iterative process. SVM and RBFNN are generally slower due to the time-consuming optimization process during the model training. Given the very high training speed, the on-line updating of the model can be readily achieved, e.g. periodically and half-daily. Once more TDS results are available, they can be attached to the original training data to re-train the IS model. For all the models, the classification/prediction process is very fast and costs around one millisecond for one single test instance, indicating a compatible decision-making speed for on-line STVS assessment.

4.11.5 Extended Accuracy Test

In the above test, only one contingency is considered. In practice, it is usually needed to assess more contingencies depending on their likelihood to occur. For a comprehensive test of the accuracy, another two contingencies are tested here. Their details are given in Table 4.14. Given the 700 generated OPs, the two contingencies are simulated, which yields another two databases. The databases are then used to train and test the hierarchical IS and compared with state-of-the-art algorithms through the same procedure of the above test. The classification and prediction performance are shown in Tables 4.15 and 4.16, respectively.

Table 4.14: Contingencies for extended test

Fault No.	Fault location	Fault duration
#2	Bus 8	0.11 s
#3	Bus 21	0.11 s

Table 4.15: Voltage collapse classification accuracy

Model	Fault #2	Fault #3
Hierarchical IS	99.2%	98.8%
Single ELM	96.7%	94.9%
SVM	97.2%	96.2%
DT	98.9%	97.4%
RBFNN	96.1%	95.3%

Table 4.16: TVSI prediction MAPE

Model	Fault #2	Fault #3
Hierarchical IS	1.66%	1.78%
Single ELM	3.03%	3.77%
SVM	2.68%	3.85%
DT	3.78%	4.43%
RBFNN	2.83%	2.69%

To summarize the overall performance, the average performance metrics on the three contingencies for each model are compared in Fig. 4.36. In average, the developed hierarchical IS still outperforms the others in both classification and prediction sub-problems, owing to the ensemble effect.

4.12 Conclusion

In this chapter, an intelligent SA framework is first presented to show how to apply IS to perform on-line SA provided today's power systems are penetrated by considerable wind energy. The framework consists of four interactive modules dedicated respectively for on-line SA, wind and load forecasting, database generation, and model updating. All the four modules are properly incorporated to accommodate the wind power and load demand variations in the system.

The essential components in designing an IS are also introduced and explained, including the data generation scheme, the feature selection process, and the machine learning algorithms. In the presented data generation scheme, variations from load demand and wind power are simulated, and OPF is performed on the extended load and wind power profiles to generate the valid OPs. The stability status of OPs is obtained via TDS subject to a set of credible contingencies. In the feature selection section, two feature selection algorithms – modified fisher discrimination criterion and RELIEFF algorithm – are introduced and used in the rest of

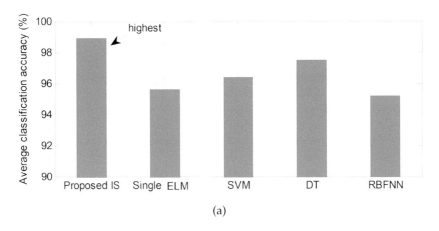

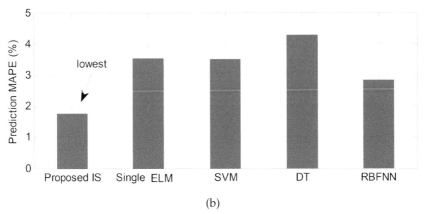

Fig. 4.36: Average performance on the three contingencies: (a) classification for voltage collapse; (b) prediction for TVSI

the book. Four types of machine learning algorithms – ANN, DT, SVM and RLA (including ELM and RVFL) – are introduced and their performances are compared using a basic IS-based SA model on New England 39-bus system. The comparison result demonstrates that, in achieving the similar level of accuracy, ELM shows the fastest learning speed, excellent generalization ability, and on-line updating/enriching ability, all of which make it an ideal solution algorithm for on-line SA task. Owing to above merits, ELM is used as the machine learning algorithm in the rest of this chapter.

By investigating and analyzing the samples misclassified by ELM-based IS, it is observed that when the ELM output is very close to the classification threshold, the determination of final class could be less reliable. Motivated by this observation, an ELM ensemble model is further developed for more accurate and more reliable SA of power systems. Under an ensemble learning scheme, the model combines a series of individual ELMs to overcome the drawbacks of the single learner and provide additional merits valuable for SA. With strategically designed training and decision-making rules, the ELM ensemble model can learn and work very fast to provide an effective estimation on the credibility of its SA results, allowing a flexible and reliable pre-disturbance SA mechanism where unreliable results can be averted for use. The parameters involved in the credible decision-making rule are further optimized under a MOP framework to best balance the tradeoff between SA accuracy and earliness. This ends up with an intelligent early-warning system for power system SA. By strategically sacrificing a negligible amount of warning accuracy, the warning earliness is significantly enhanced. Moreover, the intelligent early-warning system also enables system operators to empirically make their compromise decisions on warning model performance from an intuitive POF. The numerical testing results on New England system have comprehensively verified the efficiency, robustness, accuracy and reliability of the ELM ensemble model. In particular, the potential classification errors can be eliminated, making all the assessment decision accurate. This can be quite valuable for the intelligent early-warning system to be put into practice, since, when combined with an alternative tool such as TDS, the whole SA speed, accuracy and reliability can be enhanced to a level compatible with on-line SA requirement. The testing results also show that the flexibility of modulating the warning accuracy and earliness can make the developed system more robust to potential changes of on-line SA requirements.

Besides transient stability, STVS is another stability problem that becomes increasingly important in today's power systems with growing wind power penetration. The STVS assessment problem can be decomposed into a transient voltage collapse sub-problem, which corresponds to a classification process and unacceptable dynamic voltage deviation sub-problem, which corresponds to a regression process. TVCI and TVSI can be employed as the two indices to respectively detect voltage collapse and quantify the unacceptable dynamic voltage deviation. Based on this concept, a hierarchical IS is developed to address the two sub-problems sequentially. This IS is based on ELM ensemble model and is implemented in an off-line training, on-line application and on-line updating pattern. The hierarchical IS is validated on the New England system and compared with some other state-of-the-art learning algorithms. Simulation results show that the hierarchical IS can effectively increase the performance over a single ELM and outperform the compared algorithms in both classification and prediction sub-problems.

5

Intelligent System for Preventive Stability Control

5.1 Introduction

Along with well-established SA applications, the IS techniques have also shown encouraging potential in designing preventive control strategies owing to their ability to extract useful information about system dynamic security characteristics. Among various IS technologies, DT, as a machine learning approach, is able to provide explicit decision-making rules, attracting great interest in developing preventive control strategies. By exploiting the classification rules of the DT, control schemes against dynamic insecurities can be derived. In (E. Karapidakis and N. Hatziargyriou, 2002), a DT-based on-line preventive control of isolated power systems is presented. In (K. Mei and S.M. Rovnyak, 2004), DT is utilized for predicting one-shot stabilizing controls. In (E.M. Voumvoulakis and W.D. Hatziargyriou, 2008), load shedding strategies are inferred from inverse reading of the DT. In (K. Niazi, *et al.*, 2010), a hybrid method using NN and DT is designed for preventive generation rescheduling. In (I. Genc, *et al.*, 2010), the dynamic secure boundaries are approximated by a linear combination of DT rules, which are then applied for generation rescheduling and load shedding. Compared with conventional analytical preventive control approaches, the DT-based methods disregard the complicated system dynamics and can be much more efficient, transparent, and interpretable.

Following this promising trend, this chapter develops a new DT-based on-line preventive control strategy for transient instability prevention of power systems. Firstly, critical generators that are decisive for transient stability restoration are selected through a feature selection process based on RELIEFF. Then the generator active outputs are used as features to construct a DT. By interpreting the DT, its splitting rules regarding transient stability restoration are formulated as inequalities and are incorporated in a conventional OPF model. By solving the OPF, the preventive control can be realized as quite efficiently suitable for on-line application and the cost can be minimized. Besides, it is noticed that most of reported methods in the literature consider only one single contingency (E. Karapidakis and N. Hatziargyriou, 2002; E.M. Voumvoulakis and W.D. Hatziargyriou, 2008; I. Genc, *et al.*, 2010). While in practice, it is usually needed to include more than one disturbance in preventive control. In this chapter, the multi-contingency control is also addressed. As a case study, the developed control strategy is tested on New England 39-bus system, and its effectiveness and efficiency are demonstrated.

However, it is also known that the DT induction is a *supervised* learning process, i.e. it fully relies on the prior knowledge of the data to grow the tree. Under some particular conditions, such as class imbalance, the tree tends to be sensitive to little changes of the data. Besides, when a tree is found complex/deep in structure, some subjective judgments are needed to prune the tree such that explicit rules can be acquired for use.

Under a similar knowledge extraction and utilization framework, this chapter further applies an alternative, yet more robust statistical learning technique called Pattern Discovery (PD) (A.K.C. Wong and Y. Wang, 2003) in developing a preventive control method for transient instability prevention. PD belongs to *unsupervised* learning. In an unbiased and comprehensive manner, it statistically discovers the hidden structure in a database and provides objective, transparent, and interpretable knowledge called *patterns* for specific use (A.K.C. Wong and Y. Wang, 2003; T. Chau and A.K. Wong, 1999). The *patterns* are geometrically non-overlapped hyper-rectangles in a *Euclidean* space, easy to present and interpret. When discovered in a power system critical feature space, they can represent the dynamic secure/insecure regions, providing decision support for real-time security monitoring and situational awareness. By explicitly formulating the patterns into a standard OPF model, the preventive control can be achieved transparently and efficiently.

5.2 A Decision Tree Method for On-line Preventive Transient Stability Control

Transient stability refers to the ability of synchronous machines of an interconnected power system to remain in synchronism after being subjected to a disturbance. Its mathematical formulation for classical generator model can be expressed as follows:

$$\begin{cases} M_i \dfrac{d\omega_i}{dt} = P_{mi} - P_{ei} \\ \dfrac{d\delta_i}{dt} = \omega_i \end{cases}, i = 1, 2, \ldots, NG \qquad (5.1)$$

where for generator i, δ_i and ω_i stand for its angle and angular speed, respectively; P_{mi} and P_{ei} are its mechanical input and electrical output powers, respectively; and M_i is the machine's inertia constant.

The instability usually appears in the form of increasing angular swings, i.e. ω_i and δ_i of some generators, leading to their loss of synchronism with other generators.

As shown by equation (5.1), transient stability is strongly governed by power outputs of generators. Thus in practice, transient stability controls are mostly achieved via generation rescheduling. The procedure consists of shifting active power output from some generators to others so as to modify the OP to withstand the potentially harmful contingencies through the SA results. The key problems are the direction and the amount that generation rescheduling shall undergo.

As a machine learning technique, DT is promising to provide the generation rescheduling strategy for transient stability restoration. As recently reported by (I. Genc, *et al.*, 2010), an oblique DT is developed and the stability region is defined as the linear combination of the generator features provided by the tree. Stability control is realized by driving an unstable OP to the stable region. However, due to the high non-linear nature of the problem, the inferred stability region can be quite sensitive with respect to the variation of tree. Besides, the transparency is lost to a large extent when the tree is oblique. In this section, a more efficient and transparent strategy is proposed: given a well-developed tree, by interpreting the tree rules, active power generation can be shifted from 'unstable' terminal to 'stable' terminal along a specific branch. Taking the DT in Fig. 4.7 for example, assume an unknown OP follows along the path 'Node 1→Node 2→Terminal Node 2', yielding an 'unstable' conclusion. If the value of Node 2 of this OP can be manipulated towards 'Terminal Node 1', an alternative stability condition, 'stable', can be obtained. In doing so, the stability control can be realized. As this can be done straightforwardly and efficiently, it can be applied in on-line environment with transparency.

5.2.1 Critical Generator Identification

As already mentioned, the DT technique adopts information-based evaluation criteria for selecting tree nodes. The features with higher information contents are selected tree nodes. However, the information-based algorithms provide little indication of the distinguishing capability of features in distance space, which is important for effectiveness and economic aspect of the generation rescheduling in context of the transient stability control. Besides, there is also a strong need to use less features to construct a compact tree; otherwise, the tree could grow complicated in structure, leading to great difficulties for people to interpret.

In the literature, the trajectory sensitivity technique (T.B. Nguyen and M. Pai, 2003) and the SIME approach (D. Ruiz-Vega and M. Pavella, 2003) have been used for critical generator identification. A sensitivity analysis is performed in I. Genc, *et al.* (2010) to determine effective generation shifting pairs. These approaches only address single OP/contingency scenarios.

As previously presented in Section 4.4.2, RELIEFF is an effective method for evaluating the significance of the features with respect to the object. RELIEFF statistically evaluates the quality of features according to how well their values are distinguished among instances near each other. It not only considers the difference in features' values and classes, but also the distance between the instances; so the good attributes will make similar instances close, while dissimilar ones apart from each other. This property is very beneficial to tree growing, since homogenous instances tend to be gathered while dissimilar instances tend to be separated by good features in the distance space. Therefore, in this section, RELIEFF is adopted to identify critical generators. The critical generators are then used as features to grow a tree.

5.2.2 Generation Rescheduling within Optimal Power Flow

In order to minimize the overall cost of preventive control and to be systematic, the generation rescheduling can be realized within an OPF model.

The OPF determines the optimal control variables setting subject to an objective function and constraints. The mathematical model with minimizing generation cost can be formulated as follows:

$$\text{Minimize} \sum_{i \in S_G} (a_i + b_i P_{Gi} + c_i P_{Gi}^2) \tag{5.2}$$

$$\text{s.t.} \begin{cases} P_{Gi} - P_{Di} - V_i \sum_{j=1}^{n} V_j (G_{ij} \cos \theta_{ij} + B_{ij} \sin \theta_{ij}) = 0 \\ Q_{Gi} - Q_{Di} - V_i \sum_{j=1}^{n} V_j (G_{ij} \sin \theta_{ij} - B_{ij} \cos \theta_{ij}) = 0 \end{cases} \tag{5.3}$$

$$\underline{P}_{Gi} \le P_{Gi} \le \overline{P}_{Gi} \quad (i \in S_G) \tag{5.4}$$

$$\underline{Q}_{Gi} \le Q_{Gi} \le \overline{Q}_{Gi} \quad (i \in S_R) \tag{5.5}$$

$$\underline{V}_i \le V_i \le \overline{V}_i \quad (i \in S_B) \tag{5.6}$$

$$\underline{S}_{Li} \le S_{Li} \le \overline{S}_{Li} \quad (i \in s_L) \tag{5.7}$$

where (5.2) is the objective function, (5.3) is the power flow balance constraint set and inequality sets (5.4) to (5.7) are operational constraints; a_i, b_i, c_i are the generation cost coefficients of i-th generator, P_{Gi} is the active output of i-th generator in dispatchable generator sets S_G, Q_{Ri} is the reactive output of i-th reactive source in controllable reactive source sets S_R, V_i is the voltage of i-th bus in bus sets S_B, G_{ij} and B_{ij} are conductance and susceptance between i-th and j-th bus, respectively, and S_{Li} is the apparent power across i-th transmission line in transmission line sets S_L.

To restore the transient stability, active power outputs of critical generators can be restricted in (5.4) according to the DT splitting rules and the new OP can be obtained by solving the OPF. In doing so, preventive control can be efficiently realized with the total generation cost being minimized.

5.2.3 Computation Process

The DT splitting rules that are employed for preventive stability controls can be formulated as follows:

$$R_C = \{T \in S : N_i \leq \eta_i, i \in \vartheta\} \tag{5.8}$$

where R_C denotes a tree developed for contingency C, T denotes terminal nodes of a tree, S means 'stable' class, N_i and η_i are the node i and the corresponding threshold, respectively, and ϑ denotes the critical generator set.

The rules should be prepared in off-line via automatic tree growing and the on-line computation processed of preventive control is as follows:

(1) Perform SA on the current OP with respect to contingency C. If the OP is 'stable', stop the computation; otherwise, go to the next step;
(2) Modify tree-splitting rules (5.8) as $R_C = \{T \in S : N_i \leq d_i \times \eta_i, i \in \vartheta\}$, and set $d_i = 1$;
(3) Substitute the modified splitting rules into OPF model, i.e. substitute (5.8) into (5.4);
(4) Solve the OPF; if it converges, obtain a new OP and go to the next step; otherwise, go to step 6;
(5) Perform SA on the new OP with respect to contingency C; if it is 'stable', stop the computation; otherwise, go to the next step;
(6) Update $d_i = d_i + \varepsilon$ and return to Step 3.

Note that ε in Step 6 is a user-defined parameter; its effect is to increase the generation shifting in case the transient stability is not restored from the original tree threshold η.

5.3 Numerical Tests on Decision Tree-based Preventive Control Method

The developed DT-based preventive control method is tested on New England 10-machine 39-bus test system as previously shown in Fig. 4.22. The system data are obtained from M. Pai (2012), and the fuel cost parameters are taken from T.B. Nguyen and M. Pai (2003).

In simulation, power flow and OPF are solved using MATPOWER 4.0 package (R.D. Zimmerman, *et al.*, 2010) and SA is carried out by TDS using PSS/E software under distributed computation architecture (K. Meng, *et al.*, 2010), and DT is developed using CART 6.0 package (S. Steinberg and M. Golovnya, 2006).

5.3.1 Test System and Simulation Tools

The total active power load at the base OP of the test system is 6097.1 MW. An initial OP is obtained by solving OPF without preventive controls. The active power output of each generator is given in Table 5.1 (the upper row denotes the generator number and the lower row is the active power output), and the total generation cost is 60920.39 $/h.

Table 5.1: Generation of initial OP (MW)

G30	G31	G32	G33	G34	G35	G36	G37	G38	G39
242.7	566.4	642.2	629.8	508.1	650.6	558.2	535.3	829.4	977.0

A severe three-phase fault is assumed here (named Fault 1). The fault intervenes at bus 4 and is cleared by tripping line 4-5 after 0.3s. The TSI based on maximum rotor angle deviation (*see* Section 1.7.1) is employed to compute the transient stability degree (I. Genc, *et al.*, 2010).

For the initial OP, the TSI under this contingency is –72.88, and as shown in Fig. 5.1, it can be seen that under this disturbance, the generators lose the synchronism by diverging into different clusters. Note that in Fig. 5.1 the relative rotor angles are portrayed in COI framework. This will be also followed throughout this chapter.

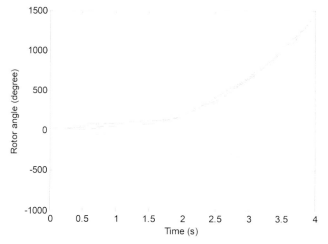

Fig. 5.1: Rotor angles swing curves of the initial OP under Fault 1 without preventive controls

5.3.2 Database Generation

To generate a stability database for training a DT, a large number of OPs should be simulated for the test system. The process consists of varying the load and generator outputs for various combinations in a neighboring range of the initial OP and producing new OPs by solving power flow under each load/generation scenario. The same contingency is imposed on the produced OPs and their TSI are calculated to indicate their stability statuses. At last, 1,000 instances are obtained in total where 231 are 'stable' and 769 are 'unstable'.

Note that in practice, the database generation should be finished in off-line before the on-line application of the method.

5.3.3 Critical Generators

RELIEFF algorithm is applied to evaluate the 10 generators of the test system, i.e. G30 to G39, and the weight of each generator is shown in Fig. 5.2.

It can be noticeably observed in Fig. 5.2 that the generators G39, G31 and G32 are positive in weight, indicating they have positive instance distinguishing capability, which means they are most decisive for the transient stability characteristic under this contingency. These three generators are selected as critical features to train the DT.

While for the other features, they are negative in weight, which means they cannot discriminate the instances in the distance space and thus are

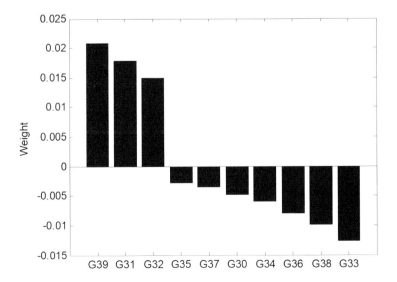

Fig. 5.2: Weight of each generator feature under Fault 1

not important to affect the transient stability of the power system under Fault 1.

In order to better illustrate the concept and show the effectiveness of RELIEFF algorithm, two two-dimensional distance spaces are plotted in Fig. 5.3, where the upper window is constituted by the first two critical features, while the lower window is constructed by the last two non-critical features. It can be seen that in the upper window, the OPs are marginally separated, while in the lower window, they are heavily overlapped. It is evident that tuning the critical features can directly alter the system stability conditions. It also verifies the excellent feature-ranking capability of RELIEFF.

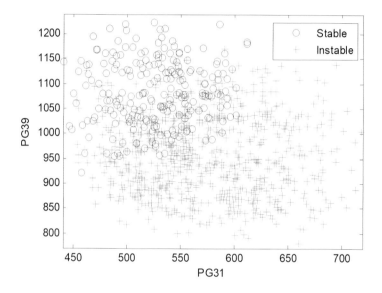

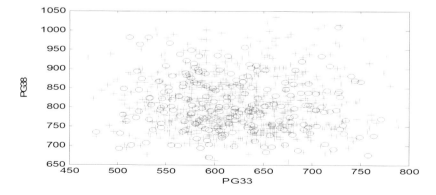

Fig. 5.3: Critical vs non-critical generator distance spaces

5.3.4 Decision Tree and Preventive Control Rules

With the critical generator features, a classification tree (CT) is developed, using algorithms integrated in CART package. *Gini* is selected as the splitting method and 10-folder cross validation criterion is selected for testing and pruning the tree. The optimal tree is shown in Fig. 5.4.

According to the tree, the rules for preventive control can be formulated as:

$$R_{\text{Fault1}} = \{PG39 > 1108.55 \ \& \ PG31 \le 597.83\} \tag{5.9}$$

It is worth indicating that the developed tree is for preventive control purpose, rather than SA which otherwise should use all the features, including other operational variables for higher classification accuracy.

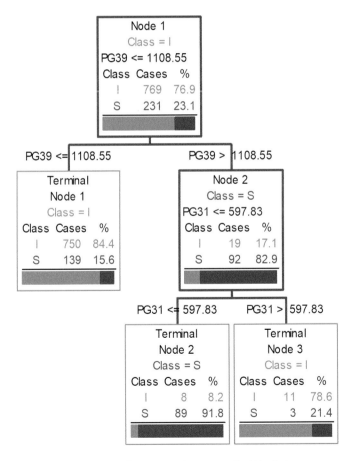

Fig. 5.4: DT for preventive control of Fault 1

5.3.5 Control Results

Following the computation process presented earlier, the developed preventive control strategy is applied on the initial OP. It is found that, after one iteration, a new, stable OP is obtained. The new OP is described in Table 5.2 and its rotor angle swing curves under Fault 1 are shown in Fig. 5.5.

Table 5.2: Generation of the new stable OP under Fault 1 (MW)

G30	G31	G32	G33	G34	G35	G36	G37	G38	G39
233.0	550.1	625.2	612.9	496.6	634.2	543.0	522.3	810.2	1108.6

As shown in Table 5.2, the critical generators' active power outputs are restricted within the DT rules; correspondingly, it can be seen from Fig. 5.5 that the new OP can maintain its stability under Fault 1, and its TSI is 41.8. The generation cost becomes 61057.51 $/h, which increases to 137.12 $/h due to the stability requirement for Fault 1.

5.3.6 Multi-contingency Control

In the literature, most of the reported methods consider only one single contingency (E.M. Voumvoulakis and W.D. Hatziargyriou, 2008; I. Genc, *et al.*, 2010), but in practice, depending on the SA results, more than one disturbance is included in preventive control. Typically, the most likely and harmful contingencies leading to insecurities should be controlled.

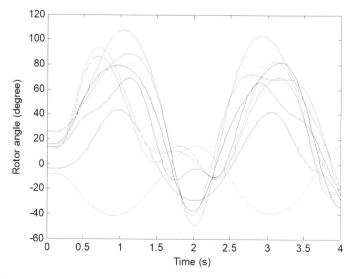

Fig. 5.5: Rotor angles swing curves under Fault 1 after preventive control

For the developed DT-based method, the multi-contingency control can be efficiently achieved by deriving separately the rules for each contingency and incorporating them simultaneously in the OPF. To illustrate this, another disturbance is assumed (named Fault 2), which is a three-phase short-circuit fault at bus 28 and is cleared by tripping line 28-29 after 0.12 s. Under this contingency, the initial OP is also unstable and its TSI is –91.6. The rotor angle swing curves under Fault 2 are shown in Fig. 5.6.

Using RELIEFF, the feature estimation results are shown in Fig. 5.7. Accordingly, G38 is identified as critical generator. The correspondingly developed DT is shown in Fig. 5.8.

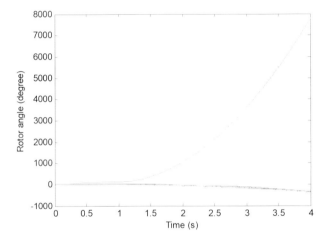

Fig. 5.6: Rotor angles swing curves of the initial OP under Fault 2 without preventive control

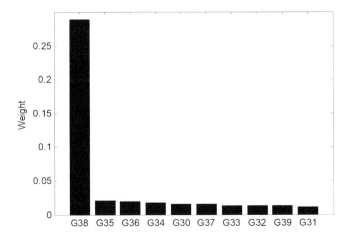

Fig. 5.7: Weight of each generator feature under Fault 2

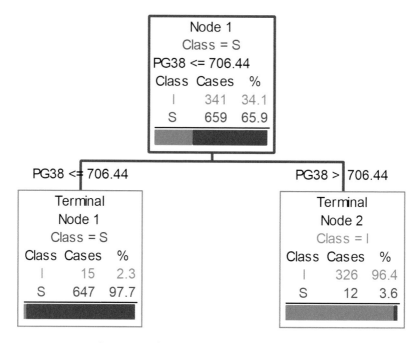

Fig. 5.8: DT for preventive control of Fault 2

It can be interpreted that the rule under this contingency can be as simple as:

$$R_{\text{Fault 2}} = \{PG38 \leq 706.44\} \tag{5.10}$$

The developed method can be quite convenient for dealing with multi-contingency. For this, the rules (5.9) and (5.10) are simultaneously incorporated in the OPF model. During the computation, SA is performed on the two contingencies, which should be both stable. After one iteration, the new, stable (under the two contingencies) OP is obtained. The generation output and costs are given in Table 5.3, and rotor angle swing curves under the two faults are shown in Fig. 5.9.

Table 5.3: Generation considering simultaneously Fault 1 and Fault 2 (MW)

G30	G31	G32	G33	G34	G35	G36	G37	G38	G39
240.5	562.7	638.6	623.0	506.9	648.8	556.6	536.8	706.4	1108.6

It can be observed from Fig. 5.9 that the new OP is able to withstand both contingencies without experiencing instability. The TSIs under

Fault 1 and Fault 2 are 38.7 and 62.4, respectively. The generation cost becomes 61160.53 $/h which increases to 240.14 $/h due to the stability requirements for the two contingencies.

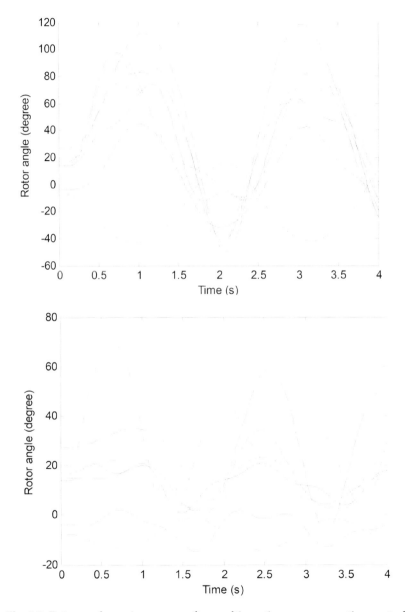

Fig. 5.9: Rotor angles swing curves after multi-contingency preventive control (upper window – Fault 1, lower window – Fault 2)

5.3.7 Concluding Remarks

The test results clearly verify the effectiveness of the developed method. Comparing conventional analytical methods, the DT-based preventive control is efficient, transparent, and interpretable.

In on-line implementation of the control strategy, the computation can be quite efficient since it is realized by solving OPF, which typically only requires several seconds (in the simulation, the OPF computation time is 0.17 s) together with a look-back SA to check if the new OP becomes stable (one SA using TDS costs 1.3 s; note that in the case study, only one iteration is needed). The major computation efforts for the method lie in the database generation, where simulating a large number of OPs and performing SA on them are required. Note that feature evaluation and DT development is quite computationally efficient. Nevertheless, all of this is carried out in the off-line stage which is not time critical. Besides, distributed computing technique can be adopted to accelerate the computation.

5.4 Pattern Discovery-based Method for Preventive Transient Stability Control

PD was proposed by Wong *et al.* (A.K.C. Wong and A.K. Wang, 1999) as an efficient knowledge extraction technique. It is unique for being able to discover non-linear and multimodal patterns of high order very fast, and rank them according to their statistical significance for interpretation, comparison, and assessment so that a greater understanding of the data can be achieved and thereby better decisions can be made. The PD has been successfully applied to numerous statistical learning problems from both academic and engineering areas, mitigating DRIP (data rich information poor) embarrassment effectively (A.K.C. Wong and A.K. Wang, 1999; T. Chau and A.K. Wong, 1999). In the previous work, PD has been employed for rule-based SA (Y. Xu, *et al.* 2010).

Based on PD technique, this section develops a preventive control method for dynamic insecurity prevention. In the off-line stage, from an SA, a critical generator feature space is first selected through a distance-based feature estimation procedure. PD is then performed in the feature space to extract patterns in an unbiased and comprehensive manner. The dynamic secure/insecure regions of the power system are represented by the unions of corresponding patterns. During the on-line implementation stage, the secure/insecure regions can provide decision support for real-time dynamic security monitoring and situational awareness. Once an insecure OP is detected through SA, the preventive control can be activated by driving the insecure OP into the secure region. To be systematic and economical, the computation process consists of formulating a pattern as explicit generator output constraints into a standard OPF model and solving the OPF. In doing so, preventive control is transparent and efficient.

To be specific, this section focuses on transient stability, but it should be noted that the methodology could also be extended for other dynamic security problems, e.g. voltage and frequency stability, and/or other control measures, e.g. discrete controls, since PD is generic not limited to a specific physical phenomenon.

5.4.1 Key Definitions in Pattern Discovery

Based on the argument that knowledge for a certain class (or group) is the significant event association inherent in the data of that class or group, PD aims to search the statistically significant subset in a *Euclidean* space in an unbiased and exhaustive manner (A.K.C.Wong and Y. Wang, 2003). Generally, the discovery process consists of residual analysis and optimization, where the former identifies significant organized information by statistically scaling the degree of the difference between actual occurrence and expected occurrence of instances, and the latter is responsible for searching out all the subtle information in the space. The fundamentals and computation procedure of PD are introduced as follows and more mathematical details and proofs could be found in (A.K.C. Wong and Y. Wang, 2003) and (T. Chau and A.K. Wong, 1999).

Consider a continuous data set Ω in the N-dimensional *Euclidean* space $\Re^N$, let $\mathbf{X} = \{X_1, X_2,..., X_N\}$ represent its feature set, and each feature X_i, $1 \leq i \leq N$, takes on values from its domain d_i, $d_i \subset \Re$, the following definitions are made for PD (A.K.C. Wong and Y. Wang, 2003; T. Chau and A.K. Wong, 1999).

Event: An event, E, is a *Borel* subset (C.M. Goldie, 1995) of $\Re^N$, while a *Borel* subset geometrically forms a N-dimensional hyper-rectangle in $\Re^N$, defined by

$$E = I_1 \times I_2 \times...\times I_N = \{\mathbf{X}: X_i \in I_i, 1 \leq i \leq N\} \qquad (5.11)$$

where $I_i = [a_i, b_i]$ is a one-dimensional semi-closed interval along the i-th feature, $-\infty < a_i < b_i < \infty$.

Volume: The volume of an event, υ, is the hyper-volume occupied by the *Borel* subset. Let L_i represents the length of the i-th interval I_i of event E, $L_i = |b_i - a_i|$, the volume of E is

$$\upsilon(E) = \prod_{i=1}^{N} L_i \qquad (5.12)$$

Observed frequency: The observed frequency of an event E, o_E, is the actual number of instances that fall inside the volume occupied by E.

Pattern: A pattern is a statistically significant event. Let $\vartheta(\cdot)$ be a test statistic corresponding to a specified discovery criterion c and θ_c^α be the critical value of the statistical test at a significant level of α. An event E is considered to be significant, i.e. a pattern, if it satisfies the condition

$$\vartheta(E) \geq \theta_c^\alpha \tag{5.13}$$

Residual: As the statistic $\vartheta(\cdot)$ to test the significance of the pattern candidates (A.K.C. Wong and Y. Wang, 2003; T. Chau and A.K. Wong, 1999), the residual of an event E is the difference between its actual occurrence, i.e. observed frequency, and its expected occurrence:

$$\delta_E = o_E - e_E \tag{5.14}$$

where e_E is the expected occurrence (frequency), under the pre-assumed model estimated by the given data set.

5.4.2 PD by Residual Analysis and Recursive Partitioning

Typically, the expected frequency e_E is estimated under uniform random distribution (T. Chau and A.K. Wong, 1999) (there are also other criteria for this (A.K.C. Wong and Y. Wang, 2003). This provides implications that if instances are randomly distributed in the space, there is no significant structure information in the space. Equivalently, the larger the residual, i.e. the difference between observed frequency and uniform random distribution, is of an event, the more structure information exists in it. Based on above definitions, PD is a process of searching all the significant events in the instance space and can be generally viewed as an optimization problem (A.K.C. Wong and Y. Wang, 2003).

In this research, a residual analysis combined with recursive partitioning procedure is adopted to discover patterns (T. Chau and A.K. Wong, 1999). The procedure consists of recursively partitioning the instance space with residual evaluation of each hyper-rectangle, until all the significant events (patterns) are identified. Its main computation steps are as follows:

(1) Divide the instance space into Q^N events, where N is the number of feature dimension and Q is the number of partitions for each feature.
(2) Refine the boundaries of each event by adjusting the event boundaries to coincide with the minimum and maximum coordinates of the contained events. This step makes the events non-overlapped on boundaries except that they have overlapped instances.
(3) Calculate the residual value of each event using the following equation (T. Chau and A.K. Wong, 1999):

$$\hat{r}_j = \frac{(n_{1j} - n_{2j})}{2 \times c_j^{1/2}} \tag{5.15}$$

where $\hat{r}_j$ is the residual of the event E_j, n_{1j} is the count of the observed contained samples of event j, n_{2j} is the expected frequency of event j, and c_j is the estimated asymptotic variance of the numerator, which can be calculated by (T. Chau and A.K. Wong, 1999):

$$c_j = \frac{1}{4}(n_{1j} + n_{2j})\left(1 - \frac{(n_{1j} + n_{2j})}{2 \times n_{1+}}\right) \tag{5.16}$$

where n_{1+} is the total count of the instances in the whole space.

(4) Evaluate the significance of each event according to the following criterion: at the $\alpha \times 100$ per cent significant level, an event is *significant* if $\hat{r}_j \geq z_{1-1/2}$; is *negative significant* if $\hat{r}_j \leq z_{1/2}$; is *insignificant* if $|\hat{r}_j| < z_{1-1/2}$, where Z_1 is the value of the standard normal deviation, $Z \sim N(0, 1)$, such that $P(Z \leq Z_1) = \alpha$.

(5) For each significant event, repeat Steps 1 to 4 to acquire more subtle description of the pattern, until termination conditions (T. Chau and A.K. Wong, 1999) are met.

Figure 5.10 shows an example of PD in a two-dimensional *sine* distribution instance space. The upper window shows the first partition and the lower window shows the last partition. It can be seen that the

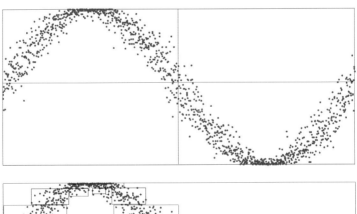

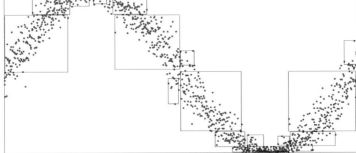

Fig. 5.10: An example of PD by residual analysis and recursive partitioning

structure knowledge in the instance space has been well identified as rectangles.

5.4.3 Database Preparation

The efficiency of the preventive control method lies strongly on the stability database where the patterns are discovered. Typically, such a database consists of a large number of instances, each associating with a pre-disturbance OP and the corresponding DSI. The OP is characterized by features, such as steady-state operating parameters; the DSI can be a discrete class label, e.g. 'secure' or 'insecure', or a continuous value, e.g. stability margin, with respect to a contingency.

To be effective, the database should reflect the possible and representative operating region that the system shall go through. In practice, the database can be acquired from historical SA archives and/or be generated via exhaustive off-line simulations. To generate a database for the upcoming period of interest, given the forecasted load levels, a range of OPs can be produced, based on the knowledge of generation scheduling information. The number of OPs required for satisfactory performance on a specific power system could be experimentally determined. Under the considered contingencies, the DSI of the produced OPs is computed through TDS. Since PD is unsupervised learning, the class imbalance and possible over-fitting problems are not the concern during the database generation stage.

5.4.4 Feature Space Selection

In the context of transient stability control, we select generator active power outputs as the features to characterize an OP. This is rational and effective because of the strong inherent coupling relationship between the generator rotor angles and the active power outputs. In the meantime, it is also necessary to select the critical generators to constitute the feature space for knowledge discovery because: (1) it is usually the case that only a subset of generators is responsible for the loss of synchronism (D. Ruiz-Vega and M. Pavella, 2003; T.B. Nguyen and M. Pai, 2003), and (2) the irrelevant generators contain very little knowledge and may be noise that hinders the knowledge extraction.

Again, this section also employs RELIEFF for selecting critical generators to compose the feature space.

5.4.5 Pattern-based Dynamic Secure/Insecure Regions

PD is performed in the critical generator feature space to extract the patterns. For each pattern, a security class label is assigned according to the possibility of different instances occurring in the pattern. For pattern E, its class label can be determined by the following rule:

$$\begin{cases} \dfrac{M_S}{M_S + M_I} > \lambda \to \text{"secure"} \\[2mm] \dfrac{M_I}{M_S + M_I} \geq \lambda \to \text{"insecure"} \end{cases} \tag{5.17}$$

where M_S and M_I respectively denote the number of 'secure' and 'insecure' instances in E, and λ is the threshold to measure the occurrence possibility of a class in a pattern (it is set as 50 per cent in this section).

The patterns, after being assigned security labels, are the structure knowledge of the power system transient stability characteristics in the critical generator feature space. The secure and insecure regions under the contingency, $\mathbf{R}^S$ and $\mathbf{R}^I$, can be represented by the unions of corresponding patterns:

$$\mathbf{R}^S = \bigcup_{i=1}^{K} E_i^S, \qquad \mathbf{R}^I = \bigcup_{j=1}^{J} E_j^I \tag{5.18}$$

where E_i^S and E_j^I denote the i-th secure and j-th insecure patterns, respectively.

Due to the explicit mathematical and geometrical forms of patterns, the secure/insecure regions can provide rule-based decision support for dynamic security monitoring and situational awareness. Typically, an OP located in the insecure region would imply an insecure status. When the dimension of the feature space is low, say two or three, the monitoring can be visualized.

5.4.6 Computation Procedure

To preventively control an insecure OP, the action involves driving it into the pattern-based secure region via generation rescheduling. Since the patterns can be mathematically formulated as a set of inequality constraints on the features, they can be seamlessly incorporated into a standard OPF model, by which the generation rescheduling can be systematically and economically determined.

An OPF model minimizing the generation costs is employed in this section. It takes the following form:

$$\min \sum_{i \in S_G} (c_{0i} + c_{1i} P_{Gi} + c_{2i} P_{Gi}^2) \tag{5.19}$$

$$\text{s.t.} \begin{cases} P_{Gi} - P_{Di} - V_i \sum_{j=1}^{n} V_j (G_{ij} \cos \theta_{ij} + B_{ij} \sin \theta_{ij}) = 0 \\[3mm] Q_{Gi} - Q_{Di} - V_i \sum_{j=1}^{n} V_j (G_{ij} \sin \theta_{ij} - B_{ij} \cos \theta_{ij}) = 0 \end{cases} \tag{5.20}$$

$$\underline{P}_{Gi} \le P_{Gi} \le \overline{P}_{Gi} \quad (i \in S_G) \tag{5.21}$$

$$\underline{Q}_{Gi} \le Q_{Gi} \le \overline{Q}_{Gi} \quad (i \in S_R) \tag{5.22}$$

$$\underline{V}_i \le V_i \le \overline{V}_i \quad (i \in S_B) \tag{5.23}$$

$$\underline{S}_{Li} \le S_{Li} \le \overline{S}_{Li} \quad (i \in S_L) \tag{5.24}$$

where c_{0i}, c_{1i}, c_{2i} are generation cost coefficients of the i-th generator, P_{Gi} is the active output of the i-th generator in the dispatchable generator set S_G, Q_{Ri} is the reactive output of the i-th reactive source in the reactive source set S_R, V_i is the voltage magnitude of the i-th bus in the bus set S_B, P_{Di} is the load of i-th bus, G_{ij} and B_{ij} are conductance and susceptance between the i-th and the j-th bus, respectively, and S_{Li} is the apparent power across the i-th branch in the branch set S_L.

Specifically, the computation of the preventive control consists of formulating the generation output limits defined by a secure pattern in (5.21) and solving (5.19)-(5.24). Note that the inclusion of a pattern into the OPF adds no complexity to the OPF model, so it can be solved by any classical programming algorithm. As the secure region may consist of many separate patterns, one needs to determine which pattern to apply in the OPF. To avoid over-control, the nearest secure pattern in the distance space could be first selected for use, but it should be noted that, due to the non-linear nature of the problem, the term 'nearest' cannot guarantee the optimality in view of OPF computations. The resulting OP may not be completely secure because of insufficient security margin provided by the nearest secure pattern. In this regard, a subsequent SA must be executed to verify the resulting OP. If the new OP is still insecure, another secure pattern should be applied. The whole computation procedure is described in Fig. 5.11.

Once an insecure OP is detected, its first-nearest secure pattern is formulated into the OPF and a new OP is obtained through the OPF solution. If the new OP is secure, the computation stops; otherwise, the next-nearest secure pattern is applied and the process repeats until a secure OP is obtained. The term 'y-th nearest' is determined by ranking the *Euclidean* distance between the OP and the centroids of the patterns. The *Euclidean* distance between two points R and R' in an N-dimensional space is calculated by:

$$D(R, R') = D(R', R) = \sqrt{\sum_{i=1}^{N}(X_i - X_i')^2} \tag{5.25}$$

where X_i and X_i' are the values of the i-th feature of instance R and R', respectively.

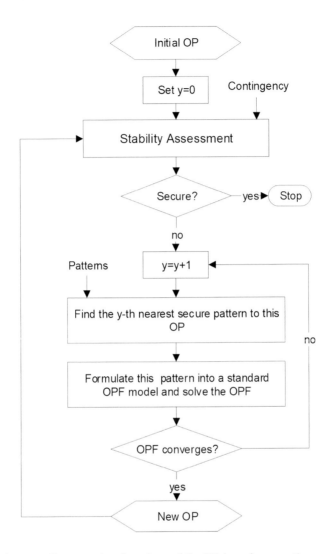

Fig. 5.11: Computation flowchart of the PD-based preventive control

The centroid of a pattern is calculated by:

$$\tau_i = \frac{X_{i1} + X_{i2} + \ldots + X_{in}}{n} \qquad (5.26)$$

where τ_i is the value of the *i*-th feature of the centroid, X_{in} is value of the *i*-th feature of the *n*-th instance in the pattern. Note that the features in distance calculation need to be normalized into a same range, e.g. [0, 1].

5.5 Numerical Tests on Pattern Discovery-based Preventive Control Method

The proposed method is tested on the New England 39-bus system. The system data is obtained from (M. Pai, 2012) and the generation cost coefficients are taken from (T.B. Nguyen and M. Pai, 2003). In the simulations, the power flow and OPF are solved using MATPOWER (R.D. Zimmeman, *et al.*, 2010), SA is performed through TDS (simulation time is 4 s), and the PD is realized in JAVA programming platform.

5.5.1 Database Generation

For the test system, its total active power load at basic operating state is 6097.1 MW. An initial OP is obtained by an OPF solution, and its total generation cost is 60920.39$/h. A three-phase fault at bus 22 cleared by tripping line 22-21 after 0.14 s is considered first (named Fault 1). SA shows that the initial OP is insecure under this fault: the generators lose synchronism (*see* Fig. 5.12) and its TSI is –90.7.

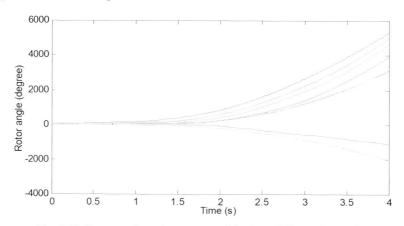

Fig. 5.12: Rotor angle swing curves of the initial OP under Fault 1

A database is artificially generated for use. The P and Q demand at each load bus are varied randomly within ±3 per cent of the initial OP; under each load distribution a set of OPs with various generation scenarios are produced in two ways: (1) randomly vary the generation outputs of each generator and solve the power flow model, and (2) randomly alter the generation cost coefficients of each generator and solve the OPF model. Note that only converged cases are saved for use. In doing so, 1,000 OPs are produced and their TSIs under Fault 1 are then obtained through SA. Figure 5.13 shows the TSI distribution of the OPs where 113 are 'secure' and 887 are 'insecure'.

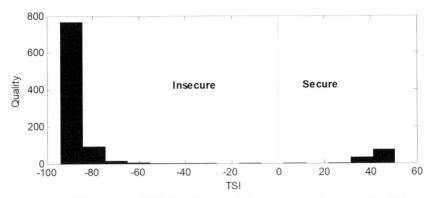

Fig. 5.13: Histogram of TSI distribution of the generated database (Fault 1)

5.5.2 Critical Generators

Regressive RELIEFF is applied to evaluate the 10 generator features. The weight of each feature is shown in Fig. 5.14, where it can be seen that generators G36 and G35 are much larger than others in weight. This indicates that they are most significant in the transient stability under Fault 1. Note that some generators have negative weights, which means that they have negative distinguishing capabilities, i.e. they are mixing rather than separating the instances in the space. G36 and G35 are selected to form the critical feature space for PD.

To better show the effectiveness of the RELIEFF algorithm, the critical feature space and a non-critical feature space constructed by G32 and G33 are visually compared. The non-critical feature space is portrayed in Fig. 5.15 and the critical feature space can be found in Fig. 5.16 (note that the patterns are also shown as the black rectangles). It can be clearly observed that in Fig. 5.15, the OPs of different classes heavily overlap, indicating

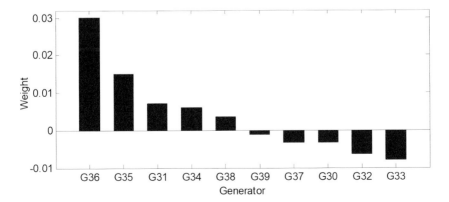

Fig. 5.14: RELIEFF feature estimation result (Fault 1)

no (or very little) structure information existing in the space. In marked contrast, the OPs in Fig. 5.16 are appreciably separated, implying evident structure knowledge about the security characteristics.

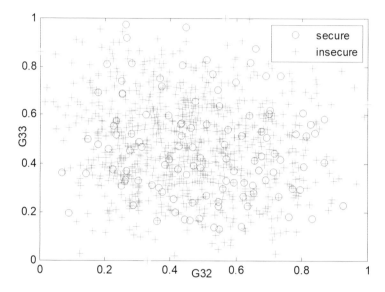

Fig. 5.15: Non-critical feature space by G32 and G33

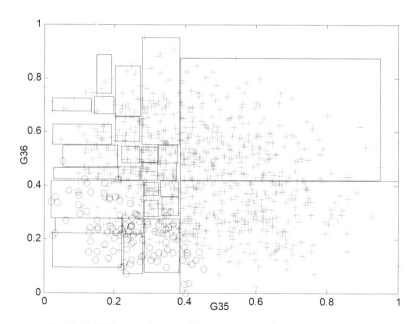

Fig. 5.16: Different classes of instances and discovered patterns

5.5.3 Pattern Discovery Results

Totally 28 events are discovered in the critical feature space where 27 of them are evaluated as significant, i.e. patterns. These 27 patterns encompass 996 instances in total and occupy 75 per cent volume of the whole instance space. As shown in Fig. 5.16, the patterns form non-overlapped rectangles in the critical generator space. It is worth mentioning that the PD is quite computationally efficient as it costs only several CPU seconds on the generated database.

The patterns are assigned security labels and their centroids are calculated and displayed in Fig. 5.17 where '+' and 'o' denote the centroids of insecure and secure patterns, respectively. Four patterns are secure and 23 are insecure.

5.5.4 Preventive Control Results

The secure/insecure regions under Fault 1 are represented by the corresponding patterns. Since the dimension of the feature space is two, it can provide an illustration of visualized dynamic security monitoring. As shown in Fig. 5.17, the initial OP (shown as the red triangle) is found located in the insecure region, which is consistent with its actual security status. To eliminate the risk of insecurity, preventive control should be armed to drive it to the secure region. Following the computation process described in Fig. 5.11, the nearest pattern of the initial OP is applied first, whose region is (0.2245, 0.2826]×(0.2223, 0.2787] in the space of G35 and G36 (normalized value). By formulating them into the OPF model and

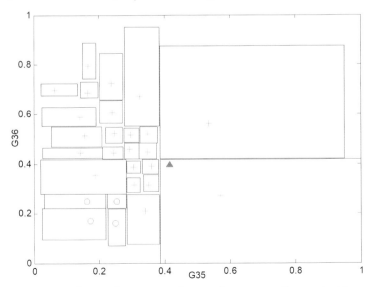

Fig. 5.17: Secure/insecure regions and corresponding centroids

solving the OPF, the first new OP is obtained. SA shows that the first new OP remains insecure (*see* Fig. 5.18). However, it is important to note that the TSI of the first new OP becomes –67.4, which has been largely improved over the initial OP. The next nearest secure pattern is then applied, yielding the second new OP, and SA shows that this OP becomes secure. Its TSI is 40.5 and its rotor angles are shown in Fig. 5.19. This new secure OP's generation cost is 61003.14$/h, which is 82.75$/h higher than the initial OP due to the control cost.

To further test the proposed method, all the 887 insecure OPs generated in the database are controlled, using the four secure patterns. It is found that all of them can be secured after the control, further confirming the effectiveness of the secure patterns. Investigation shows that, 14 OPs are secured by their first secure pattern, 737 OPs by their second nearest secure pattern, 108 OPs by their third nearest secure pattern, and 28 OPs by their fourth nearest secure pattern.

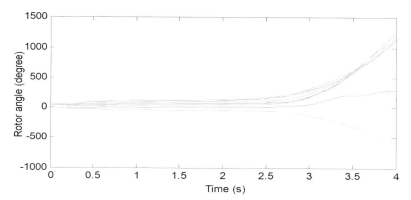

Fig. 5.18: Rotor angle swing curves of the first new OP under Fault 1

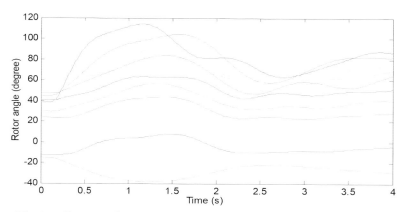

Fig. 5.19: Rotor angle swing curves of the second new OP under Fault 1

5.5.5 Multi-contingency Control

As already mentioned, most of the reported similar methods in the literature consider only one single contingency (E.M. Voumvoulakis and W.D. Hatziargyriou, 2008; I. Genc, *et al.*, 2010). Depending on the SA results, it is necessary to include more than one contingency in the preventive control. Generally, the most likely and harmful contingencies that lead to insecurities should be controlled. In this section, multi-contingency is also considered in the proposed method. The idea is to identify a general critical feature set for all the faults, and the patterns extracted in the general critical feature space can be the collective secure/insecure regions under the faults. Then the multiple faults can be simultaneously controlled in one computation run.

Another harmful disturbance is considered: it is a three-phase fault at bus 28 cleared by tripping line 28-29 after 0.10 s (named Fault 2). Under Fault 2, the TSI of the initial OP is –93.1 and for the OPs in the database, 236 are secure and 764 are insecure. A general TSI is taken as the average value of the TSIs of Fault 1 and Fault 2. It is considered that this general TSI can account for the composite security measure under the two faults. Thus the two faults can be simultaneously involved in estimating features. Regressive RELIEFF is applied to evaluate the generator features under the general TSI. According to the result in Fig. 5.20, G38, G36, and G35 should be selected as critical generators for the two faults. While G36 and G35 are critical features of Fault 1, it implies that G38 is critical to Fault 2. Interestingly, this is consistent with the feature estimation result considering only Fault 2 (*see* Fig. 5.21).

In the space of G38, G36 and G35, 22 patterns are discovered, where three are secure and 19 are insecure. The union of the secure patterns represents the collective secure region under the two faults. The initial OP is then controlled and it is found that after the first nearest secure pattern

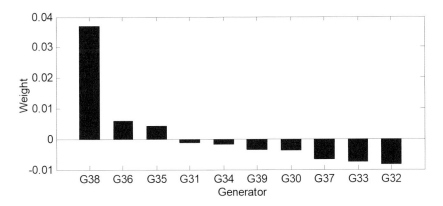

Fig. 5.20: RELIEFF feature estimation result (Fault 1 & Fault 2)

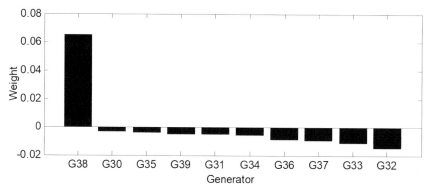

Fig. 5.21: RELIEFF feature estimation result (Fault 2)

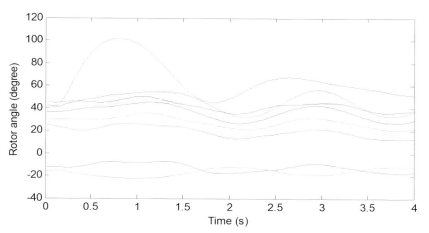

Fig. 5.22: Rotor angle swing curves of the new OP after multi-contingency control (upper window – Fault 1, lower window – Fault 2)

is applied, a new secure OP is obtained, whose TSIs under Fault 1 and Fault 2 are 45.6 and 48.2, respectively and generation cost is 61184.2$/h. The rotor angle swing curves are shown in Fig. 5.22.

5.5.6 Discussions

With the discovered patterns, the initial OP is stabilized under both single- and multi-contingency conditions, providing the advantage of interpretable and transparent control mechanism and high computation efficiency.

In the meantime, it is also clear that the success of the control relies strongly on the acquisition of secure patterns. In the case studies, due to the high severity of the two considered contingencies, most of the generated OPs are insecure, leading to only a small number of secure patterns: four for Fault 1 and three for Fault 2. In practice, efforts can be made to enrich the secure patterns, e.g. by simulating more secure OPs and/or making use of historical SA archives. This can effectively enhance the quality of the method. As the PD process is quite efficient, the database enrichment/updating can be performed on-line, using up-to-date operating information.

5.6 Conclusion

IS technologies can be applied to develop preventive control strategies. Among various IS approaches, DT is advantageous for providing explicit classification rules, which can be used to design stability controls. This chapter develops a DT-based on-line preventive control method for transient instability prevention of power system. Given a stability database, RELIEFF is applied to identify the critical generators, which are then used as features to train a DT. By interpreting the splitting rules of DT, the preventive control is realized by formulating the rules in a standard OPF model and solving it. As advantages, the developed method is transparent in control mechanism, is on-line computation compatible and convenient to deal with multi-contingency. The method is examined on New England 39-bus system considering single and multi-contingency conditions. Simulation results have verified its effectiveness and efficiency.

DT has received most research interests for developing preventive control strategies. However, it is also found that DT is usually sensitive to little changes in the database because of its supervised learning nature. As an unsupervised learning technique, PD is able to discover interpretable and unbiased knowledge from a large-scale database and has been applied for solving many real-world problems. Based on PD, this chapter develops a method for preventive dynamic security control of power systems. The idea is to extract patterns from a feature space characterized by critical generators. The patterns are a set of non-overlapping hyper-rectangles representing the dynamic secure/insecure regions of the power system. For on-line use, they can provide decision support for security monitoring and situational awareness. Preventive control involves driving the insecure OP into the secure region, which is transparent and efficient. In practice, this can be achieved systematically and economically by formulating the secure patterns as generation output constraints into an OPF model. Numerical tests are conducted on the New England 39-bus system considering both single- and multi-contingency conditions and the simulation results have verified the effectiveness of the developed method. Discussions on improving the quality of the method are also provided. As a generic knowledge discovery approach, PD can also be exploited for tackling other problems in power engineering, like poor database information.

6

Intelligent System for Real-time Stability Prediction

6.1 Introduction

As previously introduced in Chapter 2, power system SA&C proceeds in two different modes – preventive mode and emergency mode. In preventive mode, SA evaluates the stability condition of the power system with respect to a set of anticipated but not yet occurred disturbances, and the assessment results are associated with preventive control actions, such as generation rescheduling which aims to preventively modify the system's operating condition to withstand those harmful disturbances should they really occur. By contrast, in emergency mode, post-disturbance SP predicts the system stability status under an ongoing disturbance, and the prediction results are used to trigger emergency control actions, such as generator tripping, load shedding and controlled islanding, etc., which aim to impede the propagation of instability. As the previous chapters presented the ISs for pre-disturbance SA and preventive control, this chapter focuses on emergency mode and presents the ISs developed for power system post-disturbance SP.

Traditional stability analysis method is based on TDS. A major limitation of this method is the need for full and accurate information on system model, operating condition as well as the disturbance, and the requirement for extensive computational efforts. For post-disturbance power systems, the instability can occur very fast (say, within a few cycles). As such, the TDS-based method is insufficiently fast to stop the unstable propagation, and computation tools with real-time prediction capability is required. Therefore, the post-disturbance SP of power systems is also called real-time SP. In smart grid era, WAMS is envisaged as the grid-sensing backbone to enhance power grid security, reliability, and resiliency. With the increased availability of synchronous measurements provided by PMUs, IS strategy has been identified as an alternative yet promising approach for real-time SP of power systems (Z.Y. Dong, *et al.*, 2013). Compared with traditional methods, the advantages of the IS-based approach are much faster decision-making speed, less data requirement, stronger generalization capacity, and the ability to discover useful knowledge for relevant decision-support (Z.Y. Dong, *et al.*, 2013).

In real-time SP, in addition to the accuracy, a pivotal concern is the response time. The response time is the length of the time window that the IS should wait (to observe the system dynamic behavior) before making a classification on the system's stability status. To be effective, the response time should be as short as possible to allow timely emergency control actions. On the other hand, usually the case is that the longer the response time, the more system dynamic information the classifier can obtain and hence the more accurate the SP result tends to be. Consequently, there is a tradeoff between reducing the response time and maintaining a satisfactory SP accuracy. It is clear that the shorter response time the IS requires, the longer time can be left to counteract the propagation of the instability. In this aspect, there is a pressing need to further reduce the response time of real-time SP.

It is observed in the literature that the response time of most existing models is always a fixed value which means that all the post-disturbance events are classified after a pre-selected constant time window (I. Kamwa, *et al.*, 2001; I. Kamwa, *et al.*, 2010; I. Kamwa, *et al.*, 2009; N. Amjady and S.F. Majedi, 2007; N. Amjady and S. Banihashemi, 2007; A.D. Rajapakse, *et al.*, 2009; I. Genc, *et al.*, 2010; S. Rovnyak, *et al.*, 1994; E.M. Voumvoulakis and N.D. Hatziargyriou, 2010; S.M. Halpin, *et al.*, 2008; H. Bani and V.Ajjarapu, 2010; S. Dasgupta, 2014; Y. Dong, *et al.*, 2017). In fact, for different disturbances with different severity degrees, the instability may arise at different times; therefore, it is inefficient to make the decision at one constant time. Moreover, since the pre-disturbance SA is also performed on-line, it is sensible to make use of such pre-disturbance SA results to update/enrich the IS, which is, however, not well achieved in the literature due to the insufficiently fast learning/tuning speed of the learning algorithms.

This chapter aims to address the above discussed problems by developing a series of self-adaptive ISs for effective and efficient real-time SP. The different ISs are devised for different stability phenomena, including transient stability, STVS, and FIDVR. Compared with existing models, the novelty of the ISs lies in its self-adaptive decision-making mechanism, which means it can progressively adjust the response time to make the right SP decision at an appropriately earlier time. This allows the IS to provide predictive decision on powers system stability status, so that the emergency control actions can be triggered as early as possible to prevent further instability propagation. Besides, the developed ISs have a very fast learning speed to allow on-line updating with pre-disturbance SA results for performance enhancement.

6.2 Self-adaptive IS for Real-time Transient Stability Prediction

Transient stability is referred to as the ability of a power system to maintain synchronism when subjected to a large disturbance (P. Kundur, *et al.*, 2004). Transient instability can develop very fast; therefore, predicting the stability status after the disturbance clearance is essential to trigger emergency controls. This section presents a self-adaptive IS for real-time transient stability prediction (TSP).

6.2.1 Working Principle of IS in Real-Time TSP

Conventional emergency controls against transient instability are based on special protection systems (SPSs). To implement a SPS, pre-disturbance SA is performed with respect to a set of pre-defined contingencies. For unstable contingencies, corresponding remedial control actions are pre-determined and stored in a decision table. During the on-line operating stage, once a pre-defined unstable contingency is detected, the pre-determined controls will be indexed and triggered immediately. Such protections are event-based; although they are relatively fast, their implementation can be extremely complex, cumbersome and expensive (M. Begovic, *et al.*, 2005). Besides, due to the event-based nature, they can only cover a limited number of pre-defined contingencies and are applied to forecasted OPs.

In recent years, with the gradual deployment of wide-area measurement system (WAMS) in modern power systems, wide-area protection and control has gained increasing attention (M. Begovic, *et al.*, 2005). Such protection schemes are response-based, being much simpler and more accurate over the traditional event-based SPSs. To implement a response-based protection scheme, post-disturbance TSP needs to be performed to predict whether the system is going to lose synchronism following an ongoing disturbance. As illustrated in Fig. 6.1, after the fault clearance, the system dynamic behavior is observed for a certain time window and the prediction on the stability status for future time window is made. Once the system is predicted to lose stability in the near future, emergency controls, such as generator tripping, load shedding and controlled islanding, etc. will be triggered immediately.

For a fast and accurate post-disturbance TSP, an IS can be built to extract the mapping knowledge between the post-disturbance variables and the stability status. Then the TSP is converted to a very fast functional mapping procedure: $y = f(\mathbf{X})$, where $\mathbf{X}$ is the input vector corresponding to the time-series of power system dynamic variables (called features), y is the stability status and f is the extracted mapping knowledge. Owing to the fast computation speed of ISs, the post-disturbance TSP can be implemented in real-time; thus also called, real-time TSP.

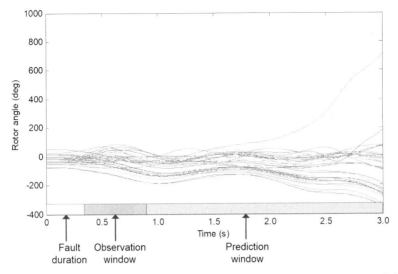

Fig. 6.1: Concept of post-disturbance TSP (the trajectories are obtained by simulating a short-circuit fault on the IEEE 50-machine system)

An illustration of such an IS is given in Fig. 6.2, which has n sets of input vector, with each set consisting of T_m variables and each variable is the value of that input at a given time point. For instance, $X_n(T_m)$ is the value of the n-th input at time T_m, y is usually a binary index, e.g. 1 and −1, denoting stable and unstable status of the system. For practical application, once the input is available, the output can be determined instantaneously. Compared with the above-mentioned analytical methods, the advantages of such an IS are much faster decision-making speed, less data requirement, and higher adaptability for different network topologies and fault scenarios. This makes it an ideal approach for real-time TSP.

It is clear that a high decision-making speed is essentially needed for real-time TSP. The TSP speed is measured by the response time, which is the elapsed time (after the fault clearance) before the IS can make a TSP classification, i.e. the time length of the observation window in Fig. 6.1 and T_m in Fig. 6.2. The response time is an essential concern since it determines the available time for the emergency controls to act. Table 6.1 reviews the response time reported in the literature.

It can be seen that the reported response time varies from four cycles to 3 s. It is usually the case that the longer the response time, the more system dynamic information the classifier can obtain and hence the more accurate the TSP result tends to be. However, since the transient instability can occur very much faster (the first-swing instability can occur as short as in two cycles), if the response time is too long, the emergence controls cannot be timely activated and thus fail to prevent the system against

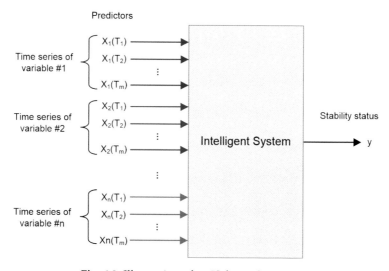

Fig. 6.2: Illustration of an IS for real-time TSP

Table 6.1: Reported response time in the literature

Reference	Response Time
[52]	2 to 3 s
[54]	1 or 2 s
[53]	150 and 300 ms
[130]	8 cycles
[55, 57]	6 cycles
[56]	5 cycles
[92]	4 cycles

instability. Consequently, there is a pressing need to further reduce the response time.

6.2.2 Feature Selection/Extraction

The input vector **X** (called features) is the synchronous time-varying variable of the power system during the post-disturbance period, sampled by PMUs. In previous works, rotor angle and angular speed have been predominantly employed (I. Kamwa, *et al.*, 2001; I. Kamwa, *et al.*, 2010; I. Kamwa, *et al.*, 2009; N. Amjady and S.F. Majedi, 2007; N. Amjady and S. Banihashemi, 2007; S. Rovnyak, *et al.*, 1994). However, these mechanical variables cannot be directly measured by the PMUs (K.E. Martin, 2015). They need to be calculated after the phasor measurement signals are transferred to the control center; hence certain time-delay and potential conversion errors are incurred.

Rather than directly using the PMU measured variables, feature extraction and selection techniques can be utilized to extract and/or select an important set of features as model inputs. A comprehensive study and comparison of feature selection methods for pre-disturbance SA is reported in Section 4.4 and (R. Zhang, *et al.*, 2012). While for post-disturbance dynamic variables, short-term Fourier transformation (STFT) (I. Kamwa, *et al.*, 2001; I. Kamwa, *et al.*, 2010; I. Kamwa, *et al.*, 2009) and empirical orthogonal functions (EOFs) (J.C. Cepeda, *et al.*, 2015) have been applied to extract composite features. The selected/extracted features can result in a lower input dimension which, on one hand, can significantly increase the training efficiency during the off-line training state and, on the other hand, require less PMU installations for real-time application. Moreover, it can contribute to improve the accuracy and allows the recognition of hidden patterns for system's stability characteristics (R. Zhang, *et al.*, 2012).

Moreover, it is worth mentioning that when using the generator dynamic variables as the model inputs, it is assumed that the PMUs are available at all generators' terminal buses. This is an ideal yet reasonable assumption because the prices of PMUs are becoming cheaper and cheaper. To be more practical, J.C. Cepeda *et al.* (2015) consider rational PMU locations based on maximum observability of slow and fast system dynamics. Using a probabilistic-based method (J.C. Cepeda, *et al.*, 2015), the best PMU monitoring locations are selected and coherent areas are defined. Then, a SVM model is trained to predict the area-based center of inertia (COI)-referred rotor angles, using the selected PMU measurements. The COI-referred rotor angles are then used to determine the stability status.

While feature extraction requires certain processing time and is relatively more complex, the authors in (A.D. Rajapakse *et al.*, 2009) proposed to directly use the post-disturbance generator voltage trajectories sampled by the PMUs as the input. As they pointed out, due to machine inertia, the generator rotor speeds and angles require longer periods of time to show a considerable deviation from their pre-disturbance steady-state values. By contrast, the generator voltage magnitudes can respond much faster than that of the rotor values to a fault. Simulation results in (A.D. Rajapakse *et al.*, 2009) have verified the effectiveness of the voltage trajectories in accurately predicting transient stability. More recently, the stronger predicting capability of generator voltage trajectories over the mechanical variables (rotor angle and angular speed) are further demonstrated through extensive tests in (F.R. Gomez, *et al.*, 2011). Following this, the generator voltage trajectories are selected as the inputs to the developed self-adaptive IS in this section. However, it should be indicated that the composite features, such as WASIs and EOF scores, can also be used for the developed IS.

6.2.3 Learning Algorithm

It is widely known that some of the learning algorithms often suffer from excessive training/tuning time; for example, it is explicitly indicated in (F.R. Gomez, *et al.*, 2011) that the tuning of a SVM classifier could cost more than 14 hours. This is because such traditional learning algorithms are based on time-consuming solutions of an optimization model or iteratively adjusting network parameters. This could impair their practical applicability. On the other hand, due to the imperfect fitting and/ or insufficient prior information in the training data, most of them can suffer from classification errors, leading to inaccurate TSP results, which should not be used directly, especially in post-disturbance environment. To enhance the training/tuning efficiency, ELM algorithm is adopted in the developed IS for extracting the mapping knowledge. Moreover, the ELM ensemble presented in Section 4.6 and the credible classification rule in Section 4.7 are employed in this section for improved classification accuracy, the ability to estimate the credibility of the classification results, and further the ability to achieve self-adaptive TSP. Besides, with its very fast learning speed, the IS can be on-line updated, using pre-disturbance SA results that can be continuously obtained on-line.

6.2.4 Self-adaptive Decision-making Mechanism

The developed self-adaptive IS organizes a series of ELM-based ensemble classifiers that have different response times. The post-disturbance TSP decision can be right made at an appropriately early time, which means the tradeoff between TSP speed and accuracy is better balanced.

The structure of the IS is described in Fig. 6.3. There are a series of individual ELM ensemble classifiers, each performing the TSP at a different response time (note that each classifier is trained using the

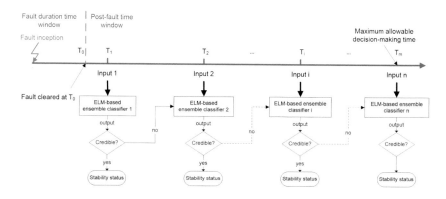

Fig. 6.3: Structure of the self-adaptive IS for real-time TSP

corresponding input vector of a response time). After the fault clearance, along the evolution of the time-varying system post-disturbance voltage trajectories, the classification on stability status will be made consecutively by the classifiers at each time point. If the classification at time T is evaluated credible, the IS will deliver the TSP result at T; otherwise, the classification will continue at time $T+1$. This procedure proceeds until a credible classification output is obtained or the maximum allowable decision-making time is reached. In practical implementation, it should be defined a maximum allowable decision-making time, e.g. $T_m = 2$ s, which means the IS must deliver a classification result when 2 s is reached. Based on such TSP mechanism, the IS can save much of unnecessary waiting time, in turn leaving more reaction time for emergency controls.

6.2.5 On-line Updating/Enriching

Due to the very fast training speed of the ELM, the TSP model can be updated on-line for performance enhancement. The updating can be done periodically, e.g. hourly or half-daily, depending on the practical needs. The data for updating can be obtained from TDS-based on-line pre-disturbance SA results. On-line pre-disturbance SA is performed in the control center and can cover a broad range of operating conditions and fault scenarios, The simulation results (including trajectories and corresponding stability status) can be saved as new training data. Such new training data can be used as a new set or attached to the initial training database to retrain the post-disturbance TSP model (*see* Fig. 6.4).

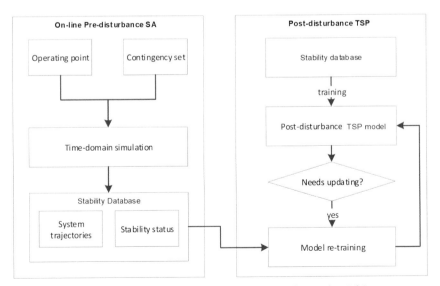

Fig. 6.4: Schematic description of on-line updating/enriching of the post-disturbance TSP model

6.3 Numerical Test for Self-adaptive TSP Model

The developed self-adaptive IS for real-time TSP is tested on two popular benchmark systems: New England 10-machine 39-bus system (F.R. Gomez, *et al.*, 2011) and IEEE 50-machine system (C.A. Jensen, *et al.*, 2001). The numerical test is conducted on a 64-bit PC with 3.1 GHz CPU and 4.0 GB RAM. The TDS is performed using commercial software PSS/E (J. Han, *et al.*, 2011), and the IS algorithms are realized in the MATLAB platform.

6.3.1 Database Generation

For each test system, 2,000 instances are generated. During the database generation, the system generation/load patterns are randomly varied within 80 per cent~120 per cent level of the initial operating condition. The corresponding OPs are obtained by running the OPF program. More details on the dataset generation procedure can be found in Section 4.3.

For each generated OP, a fault location and fault clearing time are randomly selected. To consider the probable protection delays in practice, the fault clearing time is randomly generated within 0.1 s ~ 0.3 s after the fault inception. TDS is performed to simulate the dynamics of the system, where the simulation time is 5 s and the simulation step is 0.01 s. For each OP, the simulated generator voltage trajectories and rotor angles are saved, and the maximum rotor angle TSI in (1.4) is employed to determine its transient stability condition. If the TSI is positive, the system is stable and its class label is tagged as 1; otherwise, the system is unstable and its class label is tagged as –1. The generated instances for the two test systems are summarized in Table 6.2.

Table 6.2: Generated instances

Test system	Stable instances	Unstable instances
New England	1313	687
IEEE 50-machine	1350	650

For each test system, the generated instances are then randomly divided into two sets – one serves as training set while the other as testing set, where training set occupies 75 per cent and testing set occupies 25 per cent of all the instances.

6.3.2 Model Training

To train the IS, some parameters should be properly selected, including the total ELM number E, the number of training instances per ELM d, the candidate ELM activation function ϑ_i, the optimal range of hidden node number $[h_{min}, h_{max}]$, and the credibility evaluation parameter r.

According to Section 4.9.2, E and d can be empirically set to a certain large value. $\{lb_s, ub_s, lb_u, ub_u, r\}$ should be properly set within their ranges as provided in (4.46) ~ (4.47). In this test, E is set as 200, i.e. there are 200 individual ELMs in an ensemble, d is set as 90 per cent of the total number of training instances, and r is set as 40. ϑ_i and $[h_{min}, h_{max}]$ can be optimally determined by a pre-tuning procedure, where the training set is divided into two subsets – one for training and the other for validation. The optimal hidden node range can be selected as the range which achieves the highest validation accuracy. Note that there are five initial candidate activation functions for an ELM: *'sigmoid'*, *'sine'*, *'hardlim'*, *'triangular basis'*, and *'radial basis'*. The tuning of ϑ_i is to select the best ones from them.

In this test, it is assumed that the voltage phasor measurements are sampled at a rate of one sample per cycle (0.02 s). So one ELM ensemble classifier is trained for each post-disturbance cycle, and the maximum allowable decision-making time is considered as 10 cycles (0.2 s). Hence, an IS has a total of 10 ELM ensemble classifiers trained.

The training of the IS is very fast since the ELM learning process is very computationally efficient. Table 6.3 shows the model training time for 1,500 training instances during the test. Note that since each single ELM and ELM ensemble may have different hidden node number and input number, the training time for them is varied.

Table 6.3: Training time

Model	New England system	IEEE 50-machine system
A single ELM	0.11~0.16 s	0.18~0.24 s
An ELM ensemble	24.7~30.8 s	34.7~43.2 s
The whole IS	256.8 s	415.5 s

According to Table 6.3, the developed IS has a very high training efficiency, which allows it to be on-line updated by the pre-disturbance SA results for performance enhancement.

6.3.3 Test Results

The test dataset is used to examine the trained IS. Following the self-adaptive TSP process of the developed IS, at each decision cycle, the pending (i.e. unclassified) test instances are classified by the ELM ensemble. For credible classifications, the TSP results are exported, while for incredible classifications, the test instances remain pending and move to the next decision cycle. Note that at the last decision cycle (in this test, it is 10 cycles), all the pending instances have to be classified. In this case, the ensemble TSP decision is made simply through majority voting.

The test results for the two systems are presented in Tables 6.4 and 6.5, respectively, where T_i is the *i*th decision cycle (each cycle corresponds to 0.02 s after the fault clearance); $U(T)$ is total number of unclassified instances *until* the current decision cycle; $C(T_i)$ and $C(T)$ are total number of classified instances *at* and *until* the current decision cycle, respectively; $M(T_i)$ and $M(T)$ are total number of misclassified instances *at* and *until* the current decision cycle, respectively; $A(T_i)$ and $A(T)$ are current and accumulative TSP accuracy which correspond to the classification accuracy *at* and *until* the current decision cycle, respectively, calculated as:

Table 6.4: Real-time TSP test results on the New England 39-bus system

T	U(T)	C(T$_i$)	C(T)	M(T$_i$)	M(T)	A(T$_i$)	A(T)
0	500	-	-	-	-	-	-
1	45	455	455	3	3	99.3%	99.3%
2	37	8	463	1	4	87.5%	99.1%
3	27	10	473	0	4	100%	99.2%
4	23	4	477	0	4	100%	99.2%
5	18	5	482	0	4	100%	99.2%
6	12	6	488	1	5	83.3%	99.0%
7	10	2	490	0	5	100%	99.0%
8	10	0	490	0	5	N/A	99.0%
9	7	3	493	0	5	100%	99.0%
10	0	7	500	1	6	85.7%	98.8%

Table 6.5: Real-time TSP test results on the IEEE 50-machine system

T$_i$	U(T)	C(T$_i$)	C(T)	M(T$_i$)	M(T)	A(T$_i$)	A(T)
0	500	-	-	-	-	-	-
1	161	339	339	0	0	100%	100%
2	159	2	341	0	0	100%	100%
3	148	11	352	0	0	100%	100%
4	141	7	359	0	0	100%	100%
5	120	21	380	0	0	100%	100%
6	57	63	443	1	1	98.4%	99.8%
7	49	8	451	0	1	100%	99.8%
8	44	5	456	0	1	100%	99.8%
9	25	19	475	3	4	84.2%	99.2%
10	0	25	500	3	7	88.0%	98.6%

$$A(T_i) = [C(T_i) - M(T_i)] / C(T_i) \qquad (6.1)$$

$$A(T) = [C(T) - M(T)] / C(T) \qquad (6.2)$$

According to Tables 6.4 and 6.5, it can be seen that the test instances can be classified at an earlier time with a high accuracy. Taking the New England system test for example, 455 out of 500 instances are classified at the first post-disturbance cycle, and the accuracy is as high as 99.3 per cent; the remaining 45 unclassified instances move to the second cycle, where eight of them get classified with the accuracy of 87.5 per cent (i.e. one instance is misclassified). Then the remaining 37 pending instances move to the third cycle, where 10 of them get classified with the accuracy of 100 per cent. This proceeds until the maximum allowable decision-making time (10 cycles) is reached, when all of the instances have been classified. To better illustrate the self-adaptive TSP process of the developed IS, the accumulative classification rate, calculated as $C(T)/500$, where 500 is the total number of test instances, and accumulative TSP accuracy $A(T)$ are plotted in Fig. 6.5. Here the bar chart shows the percentage of the instances that have been classified until the corresponding decision cycle and the line chart shows the corresponding TSP accuracy.

For a general estimation of the response speed and accuracy of the IS, the average response time (ART) and average TSP accuracy (ATA) can be calculated as follows:

$$ART = \sum_{i=1}^{m} [T_i \times C(T_i)] / \sum_{i=1}^{m} C(T_i) \qquad (6.3)$$

$$ATA = \sum_{i=1}^{m} [A(T)] / m \qquad (6.4)$$

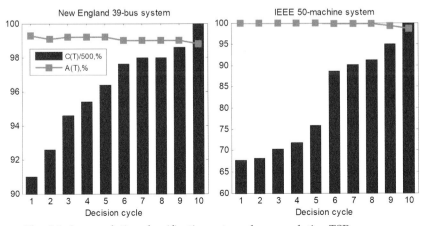

Fig. 6.5: Accumulative classification rate and accumulative TSP accuracy

where m is the total number of the maximum allowable decision-making cycles, which is 10 in this test.

The results are given in Table 6.6, where it can be seen that the decision speed of the IS is very fast and the TSP accuracy is very high. Consequently, it can allow more time for taking remedial control actions against transient instability.

Table 6.6: ART and ATA for the test

Test system	ART	ATA
New England	1.4 cycles (0.028 s)	99.1%
IEEE 50-machine	2.8 cycles (0.056 s)	99.7%

6.3.4 Updating for Performance Enhancement

The benefit of model updating/enriching is demonstrated here. In general, the more training data the model has seen, the better generalization capacity can be expected. This has already been observed in the numerical tests conducted in (Y. Xu, *et al.*, 2011). Here, we randomly generate another 500 instances for the two test systems, using the same procedure as in Section 6.3.1. The newly generated data sets are attached to the initial training datasets to retrain the IS. The updated/enriched IS is then tested on the same testing dataset. The test results are given in Table 6.7.

Comparing Tables 6.6 and 6.7, it can be seen that after the model updating/enriching, the performance improved in both terms of ART and ATA.

Table 6.7: ART and ATA for the test after
model updating/enriching

Test System	ART	ATA
New England	1.3 cycles (0.026 s)	99.4%
IEEE 50-machine	2.5 cycles (0.05 s)	99.9%

6.4 Hierarchical and Self-adaptive IS for Real-time STVS Prediction

Power system STVS refers to the ability of bus voltages to rapidly recover an acceptable equilibrium level after the occurrence of a fault (P. Kundur, *et al.*, 2004; M. Glavic, *et al.*, 2012). The leading driving force of short-term voltage instability is the inductive loads that tend to restore power consumption and draw excessive current upon a large voltage disturbance in the system (E.G. Potamiarakis and C.D. Vournas, 2006).

With the growing penetration of induction motor loads in today's power system as well as the more stringent requirement for the LVRT capacity of wind turbines, STVS problem becomes more prominent.

The recent blackout event in South Australia clearly illustrates the catastrophic impact of STVS problem on a high-level wind power penetrated power system (A.E.M. Operator, 2016). Prior to the blackout, wind power generated 51 per cent of the electricity demand. Following six successive voltage drops, nine out of 13 wind farms failed to ride through the low voltage and led to a considerable power generation loss, much faster than the UVLS scheme can react (A.E.M. Operator, 2016).

There are two different strategies to mitigate the prominent STVS problem. At planning stage, the vulnerable power grid is reinforced with dynamic VAR compensation devices, such as SVC and STATCOM (Y. Xu, *et al.*, 2014; A.S. Meliopoulos, *et al.*, 2006; J. Liu, *et al.*, 2017) , to provide permanent reactive power support. However, such devices remain very expensive, so their wide installation is usually limited by the investment budget. At post-disturbance operation and control stage, real-time STVS prediction and emergency control (e.g. load shedding) have been identified as an effective solution. In practice, for the post-disturbance voltage propagating with potential STVS problem, the timing to detect such a problem and trigger the control actions is critical to the control effectiveness. The earlier the potential problem can be detected by STVS prediction, the earlier the emergency control actions can be armed. Then there is a higher chance of the deviated voltage recovering in time. Therefore, this section aims at earlier prediction of STVS.

The STVS prediction aims to examine the time-varying dynamic voltage performance following a large disturbance. In general, the STVS problem is concerned with fast voltage collapse, sustained low voltage without recovery, and FIDVR. In the literature, most existing approaches (D. Shoup, *et al.*, 2004; S.M. Halpin, *et al.*, 2008; H. Bani and V. Ajjarapu, 2010; S. Dasgupta, 2014; Y. Dong, *et al.*, 2017; M. Glavic, *et al.*, 2012; S. Dasgupta, *et al.*, 2013; L. Zhu, *et al.*, 2015; L. Zhu, *et al.*, 2017) for real-time STVS prediction only focus on one STVS phenomenon – either voltage instability or FIDVR – but lack comprehensiveness on the full STVS problem. Therefore, it is necessary to develop an IS that can coordinate the SP on voltage instability and FIDVR phenomena. Moreover, as a post-disturbance prediction system, the prediction timing of STVS is a priority. The earlier the prediction result can be delivered, the more time can be left for emergency control actions. However, similar to TSP, most of the existing works complete STVS prediction at either a fixed time after fault clearance (S.M. Halpin, *et al.*, 2008; H. Bani and V. Ajjarapu, 2010; S. Dasgupta, 2014; Y. Dong, *et al.*, 2017), or a time very close to the voltage collapse point (D. Shoup, *et al.*, 2004; M. Glavic, *et al.*, 2012; S. Dasgupta, *et al.*, 2013; L. Zhu, *et al.*, 2015; L. Zhu, *et al.*, 2017). Such SP speed can be too slow to trigger the emergency control actions to prevent very rapid voltage-collapse events. Therefore, it is also imperative to improve the STVS prediction earliness, which refers to how fast the SP can complete.

Considering the above inadequacies, this section develops a hierarchical and self-adaptive IS to improve the comprehensiveness and earliness of real-time STVS prediction. Its main functionalities include:

(1) The detection of voltage instability and the evaluation of FIDVR severity are coordinated in a hierarchical way to improve the comprehensiveness of STVS prediction.
(2) The STVS prediction results are progressively delivered in a self-adaptive way to improve the overall SP speed.
(3) MOP is used to optimally balance the tradeoff between STVS prediction earliness and accuracy, so that the earliness can be optimized without impairing the accuracy.

6.4.1 Stability Prediction on the Different STVS Phenomena

In general, the STVS problem is concerned with fast voltage collapse, sustained low voltage without recovery, and FIDVR (M. Glavic, *et al.*, 2012; Y. Xu, *et al.*, 2015). The typical post-disturbance voltage trajectories are shown in Fig. 6.6. Figure 6.6 (a) illustrates the satisfactory post-disturbance voltage behavior in power system operation, where the voltages of all buses rapidly recover to their pre-disturbance levels after the fault is cleared. Figure 6.6 (b) shows the FIDVR scenario where some bus voltages undergo slow recovery before returning to the nominal level. Such delayed recovery may trigger UVLS and pose a significant risk to a wind turbine to ride through. Both fast voltage recovery and FIDVR scenarios are considered as stable because all bus voltages are eventually recovered.

In contrast, if the voltage cannot regain acceptable operation equilibrium, the system will be considered unstable. Such voltage instability scenario can be further classified as sustained low voltage without recovery (*see* Fig. 6.6 (c)), and fast voltage collapse – *see* Fig. 6.6 (d). Note that the fast voltage collapse is usually associated with the rotor angle instability, as Fig. 6.6 (d) shows.

In order to achieve a comprehensive STVS prediction for all the scenarios in Fig. 6.6, the STVS prediction task can be divided into two sub-tasks. The first sub-task is to decide whether the system voltage undergoes stable or unstable propagation, which is called *voltage instability detection*. If voltage instability scenario is detected, the emergency control actions should be immediately activated. However, if the system voltage is decided as stable, the FIDVR severity is then to be predicted as it serves as the second sub-task, namely, *FIDVR severity prediction*. In data-analytics area, the voltage instability detection is a time-series binary classification

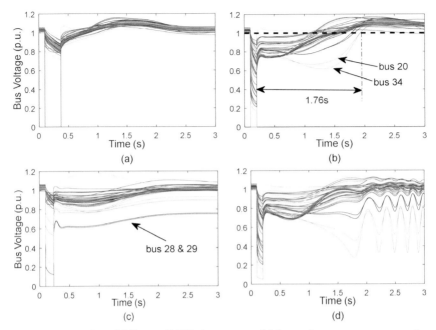

Fig. 6.6: Examples of different STVS phenomena: (a) fast voltage recovery scenario; (b) FIDVR scenario; (c) sustained low voltage without recovery (voltage instability scenario); (d) voltage collapse associated with rotor angle instability (voltage instability scenario)

problem, while the FIDVR severity prediction is a time-series regression problem.

6.4.2 Database Structure

At off-line stage, a database containing numerous post-disturbance voltage trajectory samples can be constructed through historical fault recordings and/or by running extensive TDS for a wide range of OPs, faults and load compositions. The structure of the database is demonstrated in Table 6.8 through an example. N is the number of instances, N_b is the number of buses in the system, and t_1 to t_f are the time points when PMUs take measurements. The t_f is the maximum allowable decision-making time which is the latest time when the STVS prediction result should be delivered to allow sufficient time for emergency control actions. For each instance in the database, the inputs are N_b voltage trajectories, and the two outputs are the simulated results for the two sub-tasks. For voltage instability detection sub-task, classes 1 and 0 are the class labels to respectively represent unstable and stable scenarios. For the FIDVR severity prediction sub-task, considering that FIDVR is a system-wide phenomenon, the RVSI values in (1.12) - (1.14) are adopted as the prediction target. However, if the voltage recovery performance of every single bus is concerned, VSI for each individual bus can also be used as the prediction target. Since the FIDVR severity prediction is performed only when the post-disturbance voltage is assessed as stable, the RVSI value is only available for instances in class 0.

Table 6.8: An example of STVS database

Instance ID	Bus ID	Input data (p.u.)				Outputs	
		V_{t_1}	V_{t_1}	...	V_{t_f}	Stability class	RVSI
1	1	0.46	0.48	...	0.76	1	N/A
	...	...	...	...	...		
	N_b	0.78	0.81	...	0.98		
...	...	...	...	...	...	...	...
N	1	0.82	0.83	...	0.99	0	8.8
	...	...	...	...	...		
	N_b	0.66	0.65	...	1.02		

6.4.3 Hierarchical and Self-adaptive IS

There are two major concerns regarding the real-time STVS prediction. Firstly, based on the three STVS scenarios, it is more reasonable to separate the STVS prediction task into the two sub-tasks, voltage instability detection and FIDVR severity prediction. Since the result of the voltage instability detection determines whether the FIDVR severity prediction should be carried out, the priority of the two sub-tasks is important in STVS prediction. Secondly, as STVS prediction is performed in real-time, the earlier the SP results can be delivered, the more time can be spared for emergency control actions. Thus the SP earliness (i.e. decision-making speed) should be another key aspect of STVS prediction. With the above concerns, the IS for real-time STVS prediction is developed in a hierarchical and self-adaptive architecture, as illustrated in Fig. 6.7.

In Fig. 6.7, T_c is the time point when the fault is cleared, and $\{T_1, T_2, ..., T_f\}$ are the time points when the prediction actions are executed. T_f is the maximum allowable decision-making time. As suggested in Section 4.10, the voltage instability detection sub-task (binary classification problem) is performed in the first hierarchy, while the FIDVR severity prediction sub-task (regression problem) is performed in the second hierarchy. Only the stable voltage trajectories are sent to the second hierarchy ('No' branches). Under such hierarchical structure, the two sub-tasks are prioritized and comprehensive STVS prediction is achieved.

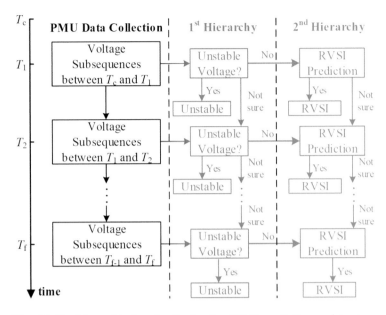

Fig. 6.7: The hierarchical and self-adaptive STVS prediction architecture of the developed IS

Moreover, in each hierarchy, the prediction proceeds in a self-adaptive way that is similar to the self-adaptive mechanism in Section 6.2.4. The outputs from the intelligent models at each time point can be either accepted or rejected by the IS. The accepted results will be directly delivered as the STVS prediction result. However, if the SP result is rejected ('Not sure' branches), the STVS prediction will be postponed to the next time point, i.e. more voltage measurement will be available. Considering that the time, when sufficiently accurate prediction results can be delivered, is naturally different for different voltage trajectories, the self-adaptive mechanism can improve the SP earliness without sacrificing the SP accuracy. The STVS prediction continues in such a hierarchical and self-adaptive way until an SP result is successfully delivered ('Yes' branches) or the maximum allowable time T_f is reached. In case T_f is reached, conventional aggregation methods, such as majority voting (for first hierarchy) or averaging (for second hierarchy) are used to ensure the successful delivery of the SP result.

At each time point, the SP is performed based on the voltage subsequences between current and the previous time points. Compared to use of the whole available voltage trajectories, use of such voltage subsequences as input can reduce the input data size, thereby reducing the computation burden of IS. Moreover, based on the voltage subsequences, critical features are selected as input for STVS prediction to further reduce the data dimension and improve the computation accuracy. The feature selection algorithm used for the IS is RELIEFF (M. Robnik-Sikonja and I. Kononenko, 2003; K. Kira and L.A. Rendell, 1992).

6.4.4 Implementation of Self-adaptive STVS Prediction

In the developed IS, separate learning models need to be trained for different hierarchies and for different time points. Classifiers are trained for the binary classification problem in the first hierarchy and predictors are trained for the regression problem in the second hierarchy. Considering that multiple classifiers/predictors are trained, the off-line training burden could be high. Therefore, ELM, benefiting from its fast learning capability (G.B. Huang, *et al.*, 2006; N.Y. Liang, *et al.*, 2006; G.B. Huang, *et al.*, 2006; G.B. Huang, *et al.*, 2011; G.B. Huang, *et al.*, 2012), is employed as the intelligent learning algorithm. Moreover, the ELM ensemble model presented in Section 4.6 is employed as each classifier/predictor in the system to achieve higher classification/prediction accuracy. The credible classification/regression rules in Section 4.7 are employed as the output aggregation strategies respectively for the first and second hierarchies. Those output aggregation strategies are used to discriminate acceptable and unacceptable ELM ensemble outputs. At on-line stage, the aggregation of unacceptable outputs will be rejected and the corresponding voltage trajectories will be re-assessed at the next time point with more effective measurement data. In doing so, the self-adaptive STVS prediction is achieved.

6.4.5 Aggregation Performance Validation

Since the output aggregation strategy is parametric, its performance is affected by the values of the aggregation parameters (i.e. $\{lb_s, ub_s, lb_u, ub_u, r\}$ for classification and $\{lb_r, ub_r, r\}$ for prediction). The performance of output aggregation can be validated through two metrics – aggregation failure rate and aggregated output error. The aggregation failure rate refers to the ratio of instances with incredible output, whereas the aggregated output error refers to the overall classification/prediction error of the ELM ensemble. In the developed IS, if the credibility requirements on the ensemble outputs are set relaxed to lower the aggregation failure rate, the overall classification/prediction error becomes inevitably larger. On the other hand, if the credibility requirements are set strict to pursue lower classification/prediction error, the aggregation failure rate becomes higher. Obviously, there is a tradeoff between the aggregation failure rate and the aggregated output error. In self-adaptive STVS prediction, lower aggregation failure rate means a lower probability of the voltage trajectories to be re-assessed at a later time point. Thus the SP can be completed earlier. Therefore, optimally balancing such tradeoff can not only minimize the aggregation error, but also optimize the SP earliness.

Previously in Section 4.8, MOP has shown its strengths in interpreting the trade-off relationship. Therefore, in this section, the tradeoff existing in the output aggregation strategy is also modeled using MOP, where the aggregation failure rate and the aggregated output error are the two independent objectives. To measure the output error, misclassification rate is used for classifiers and MAPE is used for predictors. The optimization problems for the two hierarchies are formulated as follows:

Multi-objective Programming for Hierarchical and Self-adaptive IS

By applying an ensemble consisting of D well-trained ELMs on N_v validation instances, a $N_v \times D$ regressive output set $\mathcal{O}$ is generated.

For the first hierarchy (i.e. voltage instability detection)
By applying the output aggregation strategy on $\mathcal{O}$, suppose n_s instances are successfully classified (i.e. credible output). Among them, $n_{correct}$ instances are correctly classified and n_{mis} instances are misclassified. Then the MOP problem is:

$$\text{Objectives: Minimize } [F_x(\mathcal{O}), M_x(\mathcal{O})] \tag{6.5}$$

where

$$F_x(\mathcal{O}) = \frac{N_v - n_s}{N_v} \tag{6.6}$$

$$M_x(\mathcal{O}) = \frac{n_{mis}}{n_{mis} + m_{correct}} \tag{6.7}$$

$$\mathbf{x} = [lb_s, ub_s, lb_u, ub_u, r] \tag{6.8}$$

$$\text{Constraints: } lb_s < 0, 0 < ub_s < 0.5, 0.5 < lb_u < 1, ub_u > 1, 0 < r < 1 \tag{6.9}$$

For the second hierarchy (i.e. FIDVR severity prediction)
By applying the aggregation strategy on $\mathcal{O}$, suppose the RVSI values of n_s instances are successfully predicted (i.e. credible output). The true and predicted RVSI values are respectively denoted as E_i and $\hat{E}_i$. Then the MOP problem is:

$$\text{Objectives: Minimize } [F_x(\mathcal{O}), MAPE_x(\mathcal{O})] \tag{6.10}$$

where

$$MAPE_x(\mathcal{O}) = \frac{1}{n_s} \sum_{i=1}^{n_s} \left| \frac{\hat{E}_i - E_i}{E_i} \right| \tag{6.11}$$

$$\mathbf{x} = [lb_r, ub_r, r] \tag{6.12}$$

$$\text{Constraints: } 0 < lb_r < 1, ub_r > 1, 0 < r < 1 \tag{6.13}$$

To find the optimal result for the MOP problems, an aggregation performance validation process is performed after all the ensemble learning models are trained. Since the sub-tasks in the two hierarchies are different and the effectiveness of the training features at each time point varies, the performances of the ELM ensembles trained for each hierarchy and at each time point should be different. Therefore, aggregation performance validation is performed separately for each ELM ensemble and the best set of aggregation parameters is optimized for each ELM ensemble.

The aggregation performance validation process for an ELM ensemble is shown in Fig. 6.8. From the original database, a small portion of the instances is randomly selected to form the validation dataset. By applying ELM ensemble to the validation dataset, the regressive outputs from single ELMs in the ensemble are collected. In the optimization, $M_x(\mathcal{O})$ or $MAPE_x(\mathcal{O})$ are computed, based on the difference between aggregated outputs and the simulated outputs of the validation instances. Different from a single-objective optimization problem, MOP has two or more objectives simultaneously and a set of equally optimized solutions can be derived (K. Miettinen, 2012). The final selection of the best parameter set from the Pareto set is made, based on the practical needs of decision makers.

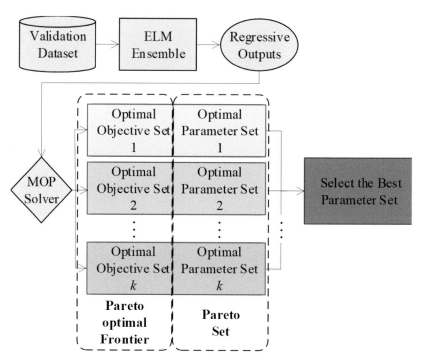

Fig. 6.8: The aggregation performance validation process

6.5 Numerical Test for Hierarchical and Self-adaptive IS

The developed hierarchical and self-adaptive IS is tested on New England 39-bus system. Considering the effect of wind power generation on STVS, the power plant on bus 37 is replaced by a 700 MW wind farm.

6.5.1 Database Generation

To obtain a comprehensive database for STVS prediction, the following factors are considered in the simulation:

(1) *Pre-disturbance OP*
    The post-disturbance voltage behavior depends on the pre-disturbance OP of the system. Using Monte Carlo simulation, 10,000 operating scenarios are sampled, based on the variation of wind power and load demand. The power generated from the wind farm is randomly sampled between 0 and its capacity 700 MW, while the load on each load bus is randomly sampled between 0.8 and 1.2 of its power rating. All other pre-disturbance quantities are dispatched by AC OPF. After removing the divergent scenarios, eventually, 9,932 OPs are subject to TDS to examine STVS.

(2) *Dynamic load modeling*
    An industry-standard composite load model CLOD (P. Sieman, 2011) defined in PSS/E is employed to build various load compositions. The six load types included in CLOD are the LM, SM, discharge lighting, transformer exciting current, constant power loads, and voltage dependent loads, all of which are typical load components in real-world substations. In the test, K_P is selected as two to simulate the voltage-dependent load, and the penetration of induction motor load is randomly selected between 0 per cent and 80 per cent for each operating scenario, so as to comprehensively simulate the impact of induction motor loads on STVS. In practice, the different load components' percentage share can be obtained by measurement-based load modeling approaches (R. Zhang, *et al.*, 2016).

(3) *Fault types and fault duration*
    A fault set consisting of 180 $N - 1$ faults is created. The faults are three-phase faults on buses or transmission lines, either cleared without loss of element or cleared by single line tripping. In doing so, the faults with and without system topology changes are considered. The fault for each OP is randomly selected from the fault set with randomly assigned fault duration between 0.1 and 0.3 second.

(4) *Simulation running and database generation*
    The TDS is run in TSAT (M. Robnik-Sikonja and I. Kononenko, 2003). The simulation step size is set to 0.01 second, which is comparable to

the PMU measurement rate. The maximum simulation time is five seconds which is considered sufficient for STVS analysis. Based on whether all bus voltages successfully recover to their pre-disturbance levels, 4,926 out of the 9,932 instances are unstable (class 1) and the rest 5,006 instances are stable (class 0). The RVSI values of stable instances are computed as the prediction target in the second hierarchy. The 9,932 instances are randomly divided into training, validation, and testing instances in a ratio of 3:1:1.

6.5.2 Feature Selection

Since the real-time measurement rate is very high, the voltage difference between consecutive measurements is considered insignificant to make improvement in STVS prediction accuracy. Therefore, in the test, voltage instability detection and FIDVR severity prediction are performed every 0.04 second. The maximum allowable decision-making time is set at 0.8 second. Thus a complete STVS prediction needs 20 (i.e. 0.8/0.04) classifiers and 20 predictors. Note that the 0.8 second maximum allowable decision-making time is an empirically selected value in this test, and, in real applications, it can be adjusted to be longer or shorter, depending on the practical requirements of SP accuracy and emergency control timing. In general, a longer maximum allowable decision-making time can increase the overall SP accuracy, but will reduce the time left for emergency control.

RELIEFF is applied for further feature selection. In Fig. 6.9, the RELIEFF results on the classification data at the first and the twentieth time points are shown as examples, where the first 101 and 129 features are selected as the training features respectively.

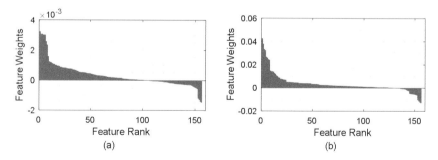

Fig. 6.9: RELIEFF results on the classification data at (a) the first time point; (b) the twentieth time point

6.5.3 Model Training

In the study, each ELM ensemble consists of 200 ELMs and each single ELM classifier/predictor is trained by 4,000/2,000 training instances respectively. The *sigmoid* function is used as the activation function due to its highest training accuracy. The only parameter to be pre-tuned is the optimal hidden node range for each ELM ensemble. The tuning results of the first and the twentieth ELM ensemble classifiers are presented as examples in Fig. 6.10, and [800, 1,000] is selected as the optimal hidden node range for the first ELM ensemble classifier while [500, 700] is for the twentieth ELM ensemble classifier.

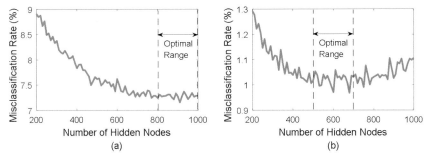

Fig. 6.10: Hidden nodes tuning results for (a) the first ELM ensemble classifier; (b) the twentieth ELM ensemble classifier

6.5.4 Aggregation Performance Validation Results

Based on the validation instances, the MOP problems (6.5)-(6.12) are solved using NSGA-II. A POF is derived for each ELM ensemble. Figure 6.11 shows the POF of the first and the nineteenth ELM ensemble classifiers as examples. The tradeoff between the aggregation failure rate and the misclassification rate can be clearly observed.

Among the Pareto points, a selection criterion is defined to select the best objective point. For voltage instability detection, the misclassification rate target is set at 0 per cent (minimum) for the first 19 ELM ensemble classifiers and the Pareto point achieving that target is then the best objective point (e.g. point *A* and point *B* in Fig. 6.11). Following the above

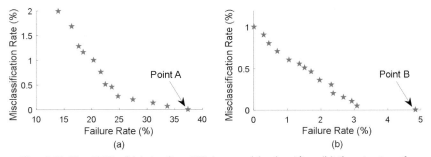

Fig. 6.11: The POF of (a) the first ELM ensemble classifier; (b) the nineteenth ELM ensemble classifier

selection criteria, the misclassification rate is minimized for the first 19 predictions, and a termination is guaranteed at the twentieth prediction by using majority voting.

For FIDVR severity prediction, the selection criterion is similar to voltage instability detection. The only difference is that for such a regression problem, zero MAPE can never be achieved. Therefore, in this test, the MAPE target is set at two per cent for the first 19 predictions and the Pareto point with MAPE just below that target is selected as the best one.

6.5.5 Off-line Computation Efficiency

In the developed IS, off-line computation includes ELM ensemble training and aggregation performance validation. The computation is performed on a PC with i7-6700 3.4 GHz CPU and 16 GB RAM. With such hardware, the consumed off-line computation time is shown in Table 6.9 where it takes 1,100 seconds to implement the ELM ensemble training and 950 seconds to validate the aggregation performance. Based on this high off-line computation efficiency, if the training database is outdated (e.g. after several days' use), the model can be re-trained efficiently using the updated data.

Table 6.9: Off-line computational time (CPU time)

ELM ensemble training	Aggregation performance validation	Total
1100 s	950 s	2050 s

6.5.6 Testing Results of Real-time STVS Prediction

There are 1,987 testing instances, including 984 unstable instances and 1,003 stable instances. The developed IS is tested on these instances under the hierarchical and self-adaptive prediction architecture. The testing results are presented in Table 6.10 and the involved nomenclature is presented in Table 6.11.

In the first hierarchy, more than half of the testing instances are successfully classified in 0.08 second (first and second time points) and 82.39 per cent of the testing instances are successfully classified in 0.2 seconds (first and fifth time points), indicating the high STVS prediction speed of the developed IS. Besides, for most time points before the twentieth prediction, the misclassification rates achieve 0 per cent, which matches with the target accuracy in the aggregation performance validation process. The overall misclassification rate is as low as 0.91 per cent, demonstrating high accuracy in voltage instability detection sub-

Table 6.10: STVS prediction testing results

T_i	First hierarchy					Second hierarchy		
	Voltage instability detection					FIDVR severity prediction		
	$R_c(T_i)$	$S_c(T_i)$	$L_1(T_i)$	$L_0(T_i)$	$E_c(T_i)$	$R_r(T_i)$	$S_r(T_i)$	$E_r(T_i)$
1	1987	761	485	276	0%	276	0	N/A
2	1226	348	100	248	0.57%	524	0	N/A
3	878	204	91	113	0%	637	0	N/A
4	674	125	102	23	0%	660	0	N/A
5	549	199	144	55	1.51%	715	22	2.2%
6	350	49	13	36	0%	729	185	2.1%
7	301	24	3	21	0%	565	138	1.8%
8	277	9	0	9	0%	436	288	2.1%
9	268	11	3	8	0%	156	74	2.2%
10	257	19	4	15	0%	97	25	2.5%
11	238	19	2	17	0%	89	17	2.3%
12	219	14	10	4	0%	76	26	2.1%
13	205	8	1	7	0%	57	30	1.5%
14	197	35	4	31	0%	58	7	2.6%
15	162	35	8	27	2.86%	78	33	2.2%
16	127	12	3	9	0%	54	19	2.4%
17	115	16	2	14	0%	49	29	1.8%
18	99	5	0	5	0%	25	3	2.0%
19	94	28	2	26	0%	48	28	2.0%
20	66	66	15	51	12.1%	71	71	7.5%
Overall	1987	992	995		0.91%	N/A	995	2.4%

Table 6.11: Nomenclature in Table 6.10

T_i	Time points.
R_c	The number of instances to be classified in the first hierarchy.
S_c	The number of successfully classified instances.
L_1	The number of instances classified as unstable.
L_0	The number of instances classified as stable.
E_c	Misclassification rate.
R_r	The number of instances to be predicted in the second hierarchy.
S_r	The number of successfully predicted instances.
E_r	MAPE.

task. In the second hierarchy, no testing instance can be successfully assessed in the first four predictions because no Pareto point with MAPE lower than two per cent can be found in their POFs. All the instances are accumulated to later time points and the first successful prediction occurs at the fifth time point. Despite the delay at the beginning, the RVSI values of more than half of the stable instances are successfully predicted before the ninth time points. The MAPE of the first to the nineteenth predictions are kept close to the two per cent target level and the overall MAPE is 2.4 per cent. In statistics, the average result delivery time for voltage instability detection and FIDVR severity prediction is only 0.17 and 0.39 second, respectively. Such short prediction time in both hierarchies verifies the exceptional early STVS prediction ability of the developed IS.

Another aspect of the STVS prediction performance to be considered is how early the unstable instances can be successfully detected in the first hierarchy, which decides how much time can be left for the emergency control actions. We use the following earliness rate to quantify the STVS prediction earliness:

$$E = 1 - \frac{t_{\text{detect}}}{t_{\text{collapse}}} \times 100\% \tag{6.14}$$

where E represents the earliness rate; t_{detect} is the time needed to successfully detect the voltage instability event after the fault is cleared; and t_{collapse} is the time elapsed since fault clearance till when the voltage of any bus in the system reaches its local minimum and remains at low level. Higher earliness rate is favorable in power system transient analysis because it indicates more time left for emergency control actions. In this test, the average earliness rate over all unstable testing instances is 61.2 per cent, meaning a considerable amount of time is spared for subsequent emergency control actions.

6.5.7 Comparison with Fixed-time STVS Prediction

By using the developed IS, STVS is assessed in a self-adaptive way to improve SP earliness without impairing the SP accuracy. It is necessary to compare the SP performance of the developed IS with fixed-time SP (i.e. assess the post-disturbance voltage trajectories at a fixed time after the fault clearance) to demonstrate its advantage. For the fixed-time SP, by applying majority voting on each ELM ensemble classifier, the voltage instability detection decisions on all testing instances can be deliberately delivered at each time point without any self-adaptive process. The misclassification rates of voltage instability detection at different time points are listed in Table 6.12. Note that only the misclassification rates at the eighteenth, nineteenth and twentieth time points are comparable to the misclassification rate in self-adaptive SP (i.e. 0.91 per cent), but the misclassification rates at other time points are much higher, meaning that the fixed-time SP must be carried out no earlier than 0.72 second (i.e. the eighteenth time point) to achieve the same level of accuracy. By contrast, the developed IS delivers the voltage instability-detection decision much earlier (i.e. 0.17 second on average), which can be 4.24 times faster. This exceptional fast STVS speed is very important at the post-disturbance stage since it can trigger very early emergency control actions to protect the power system from instability.

Table 6.12: Fixed-time STVS prediction performance

Time points	1	2	3	4	5	6	7	8	9	10
$E_c(T_i)$ (%)	6.6	6.5	4.0	2.5	2.2	2.1	2.1	1.6	1.4	1.5
Time points	11	12	13	14	15	16	17	18	19	20
$E_c(T_i)$ (%)	1.3	1.4	1.3	1.3	1.3	1.2	1.1	0.9	0.9	0.9

6.5.8 Comparison with Non-hierarchical STVS Prediction

By using the developed IS, STVS is assessed hierarchically to achieve comprehensiveness. So it is also necessary to compare the developed IS with the traditional non-hierarchical STVS prediction. Rule-based industrial criterion (D. Shoup, *et al.*, 2004) is used in the test to represent a non-hierarchical STVS prediction strategy (i.e. the voltage trajectory samples are labeled as unstable if any voltage dip exceeds a voltage magnitude threshold, V_{min}, for an unacceptable length of time, T_{max}, otherwise stable), and the self-adaptive mechanism remains in this STVS prediction test. In this test, V_{min} is 0.9 p.u. and T_{max} is 0.5 second, which is regarded as a practical STVS requirement according to (D. Shoup, *et al.*, 2004). As a result, the overall misclassification rate of non-hierarchical STVS prediction is 2.42 per cent, which is significantly higher than the 0.99 per cent misclassification rate of the developed IS. Therefore, the STVS prediction, using the developed IS, is 2.44 times more accurate than the non-hierarchical STVS prediction.

The reason for such depreciated accuracy for the non-hierarchical STVS prediction could be the increase in the number of marginal testing instances. By using the quantitative decision-making boundary defined by V_{min} and T_{max} to distinguish stable and unstable instances, there will be a large number of marginal instances that are very close to the boundary. The classification on those marginal instances can hardly be accurate, using such a simple method. However, by using the developed IS, the unstable instances are first defined qualitatively based on the pattern of post-disturbance voltage, which can greatly reduce the chance of marginal instances. Thus significantly higher accuracy can be achieved. Therefore, compared to direct binary classification based on the industrial criterion, the developed IS detects the voltage instability scenario in the first hierarchy, and then quantitatively predicts the FIDVR severity in the second hierarchy to achieve more refined STVS prediction result.

6.6 Hybrid Self-adaptive IS for Real-time STVS Prediction

Under the self-adaptive scheme, ensemble learning has been adopted to evaluate the credibility of the SP results. As an emergent learning technique, RLA, including ELM and RVFL, have shown their advantages in constructing the ensemble model, owing to their stochastic, fast, and diversified learning capability. Nevertheless, the ensemble models in Sections 6.2 and 6.4 only use ELM as a single learning unit. Considering the unique advantages of different RLAs, a single learning algorithm may not fully map the relationships embedded in the training data, therefore leading to sub-optimal STVS prediction performance.

Focusing on the STVS prediction problems, this section upgrades the self-adaptive IS in Sections 6.2 and 6.4 into a hybrid self-adaptive IS, aiming to optimize both the SP accuracy and speed. Rather than using a single learning algorithm, this hybrid IS incorporates multiple RLAs in an ensemble form, namely, a hybrid randomized ensemble model, to obtain a more diversified machine learning outcome. Moreover, under a modified MOP framework, the self-adaptive mechanism of hybrid randomized ensemble model can also be optimized to achieve the best balance between STVS accuracy and speed. The hybrid self-adaptive IS is distinctive from the self-adaptive IS in Sections 6.2 and 6.4 as follows:

(1) The hybrid randomized ensemble model consists of multiple RLAs to improve the learning diversity. Their aggregate output is shown to be more robust and more accurate than a single learning algorithm.
(2) The parameters involved in each time window together with the participation rate of each learning algorithm are optimized by solving only one MOP problem for all time points to pursue the optimal balance between SP speed and accuracy. By offering a set of optimal solutions to the system operators, the performance of the hybird IS can be adjusted according to different utility requirements.

6.6.1 Motivation of Hybrid Learning

Ensemble learning has been recognized as an effective learning paradigm to provide credibility estimation of the model outputs. In the literature, the existing ensemble learning models used for power system SA are based on a single learning algorithm, such as SVM ensembles (M. He, *et al.*, 2013), DT ensembles (I. Kamwa, *et al.*, 2010; T.Y. Guo and J.V. Milanovic, 2013), ANN ensembles (Z. Guo, *et al.*, 2012), etc. The ensemble of a single algorithm may not be diverse enough and therefore suffers from an accuracy bottleneck. One strategy to overcome such bottleneck is to integrate multiple RLAs, e.g. ELM and RVFL, into a new hybrid ensemble model. By doing so, the aggregated output error tends to decrease owing to the compensation between different learning algorithms.

A preliminary test has been carried out on two different STVS datasets to demonstrate the effectiveness of the hybrid ensemble model in improving machine learning performance. The two different datasets consist of 3,000 and 5,000 instances with post-disturbance voltage trajectories as the model input and the simulated voltage stability state as the model output. In this test, the classification performance of the ensemble models of different machine learning algorithms (including ANN, SVM, DT, ELM and RVFL) and a hybrid ensemble model incorporating ELM and RVFL are compared in Fig. 6.12. The preliminary testing results demonstrate the accuracy advantage of RLAs (i.e. ELM and RVFL) in ensemble learning and further accuracy enhancement by hybridizing ELM and RVFL. Although the hybrid ensemble model is a combination of ELM and RVFL, its overall accuracy is higher than the ensemble of either ELM or RVFL only. This means incorporation of multiple RLAs in an ensemble model tends to achieve overwhelmingly higher accuracy than whichever single algorithm with the best accuracy. When the hybrid ensemble model is deployed under the self-adaptive decision-making scheme, its advantage in accuracy could be transformed into two dimensions (i.e. accuracy and speed), which are both important performance indices in real-time STVS prediction.

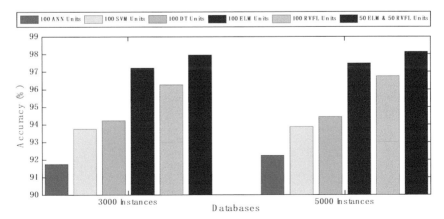

Fig. 6.12: Performance comparison of different ensemble models

6.6.2 Hybrid Randomized Ensemble Model

Based on the above positive results from the preliminary test, it is verified that a hybrid ensemble model should achieve better STVS prediction performance as compared to models with a single learning algorithm. Therefore, a hybrid randomized ensemble model is designed for self-adaptive STVS prediction. It combines the two different RLAs (i.e. ELM and RVFL) as the single learning unit, which can, not only increase the accuracy, but also provide a potential classification and prediction error to allow for more flexible and reliable mechanism before failure.

The training procedure for each single learning unit of the hybrid randomized ensemble model is shown as Algorithm 1.

Algorithm 1: Hybrid Randomized Ensemble Model

Initialize
Decide the total number of single learning units (including m ELMs and n RVFLs ($m + n = E$))
Decide the effective activation function.

Randomly choose instances **s** from the original database **S.**

Randomly choose features **f** from the original feature set **F.**

Train m ELMs

for i = 1 to m do:
Randomly assign ELM hidden layer nodes h_E within the optimal range $[h_{min}, h_{max}]$ (subject to a tuning process).
Train the ELM with the selected instances, features, number of hidden layer nodes, and activation function.
end for

Train n RVFLs

for j = 1 to n do:
Randomly assign RVFL hidden layer nodes h_R within specific optimal range $[h_{min}, h_{max}]$ (subject to a pre-tuning process).

Train the RVFL with the selected instances, features, number of hidden layer nodes, and activation function.

Constitute the Hybrid Randomized Ensemble Model

Combine m ELMs and n RVFLs in one ensemble model.

In the hybrid randomized ensemble model, the reason for adopting RLAs is twofold:

(1) Since ELM and RVFL both randomly assign input weights and biases, their training speed is highly fast, which apparently alleviates the increased computation burden of training a large number of single learning units in an ensemble model.

(2) The key to a successful ensemble model is that the adopted single learning units increase the learning diversity (Y. Ren, *et al.*, 2016). The stochastic nature of the RLAs entails further diversified learning outcome.

6.6.3 Optimal Accuracy-speed Balancing

As already mentioned, SP accuracy and speed conflict with each other, and there should be a tradeoff between them. In the hybrid self-adaptive IS, the trade-off relationship is modeled as a MOP, based on which the participation rate and the credible decision parameters are optimized to achieve the optimal STVS prediction speed and accuracy.

The MOP for optimally balancing the SP accuracy and speed is formulated as follows:

$$\underset{\mathbf{x_c}, \mathbf{x_f}}{Maximize} \quad \mathbf{p}(\mathbf{x_c}, \mathbf{x_f}) = [p_1(\mathbf{x_c}, \mathbf{x_f}), p_2(\mathbf{x_c}, \mathbf{x_f})] = [Accuracy, Speed] \quad (6.15)$$

where

$$\mathbf{x_c} = [\mathbf{lb_u}, \mathbf{ub_u}, \mathbf{lb_s}, \mathbf{ub_s}, \mathbf{r}]$$

$$\mathbf{x_f} = [\mathbf{M}, \mathbf{N}]$$

$$\begin{cases} \mathbf{lb_u} = (lb_u^1, lb_u^2, lb_u^3, ..., lb_u^{T_{max}}) \\ \mathbf{ub_u} = (ub_u^1, ub_u^2, ub_u^3, ..., ub_u^{T_{max}}) \\ \mathbf{lb_s} = (lb_s^1, lb_s^2, lb_s^3, ..., lb_s^{T_{max}}) \\ \mathbf{ub_s} = (ub_s^1, ub_s^2, ub_s^3, ..., ub_s^{T_{max}}) \\ \mathbf{r} = (r^1, r^2, r^3, ..., r^{T_{max}}) \\ \mathbf{M} = (m^1, m^2, m^3, ..., m^{T_{max}}) \\ \mathbf{N} = (n^1, n^2, n^3, ..., n^{T_{max}}) \end{cases} \quad (6.16)$$

In (6.15), the SP speed and accuracy are the two objectives relating to the participation rate $\mathbf{x_f}$ and the credible decision parameters $\mathbf{x_c}$ through p_1 and p_2, respectively. $\mathbf{M}$ and $\mathbf{N}$ represent the quantity of ELM classifiers and RVFL classifiers for each decision cycle, respectively. Based on the self-adaptive scheme, the quantity of optimization parameters depends upon the maximum allowable decision-making time T_{max}. Besides, as mentioned in Section 6.4.5, there are bounding constraints on the elements in $\mathbf{x_c}$ and $\mathbf{x_f}$. Overall, the constraints for the MOP are as follows:

$$S.t. \quad \begin{cases} lb_u^k < -1, \ -1 < ub_u^k < 0, \ 0 < lb_s^k < 1, \ ub_s^k > 1 \\ 0 < R^n < E \\ m^k + n^k = E \\ 1 \le k \le T_{\max} \end{cases} \quad (6.17)$$

Compared with the MOP in Section 6.4.5, the difference of the above MOP model is that only one single MOP is formulated to optimize the parameters for all the time points as a whole. Moreover, the objectives of the MOP in Section 6.4.5 are the failure rate and misclassification rate, while the objectives above are average decision accuracy and average decision speed to directly balance the accuracy and speed for post-disturbance STVS prediction purpose. Similar to the POF presented in previous sections, the POF here can also provide the system operators greater selection to make their compromise decision according to their actual needs.

6.6.4 Implementation of the Hybrid Self-adaptive IS

The hybrid self-adaptive IS integrates the hybrid randomized ensemble model, the credible classification rule, and MOP to achieve optimized self-adaptive STVS prediction. Its implementation requires two stages – an off-line stage and an on-line stage, which is illustrated in Fig. 6.13.

At the off-line training stage, a hybrid randomized ensemble model should be trained for each time point, and its associated participation rate and credibility checking rule are also optimized by solving the MOP

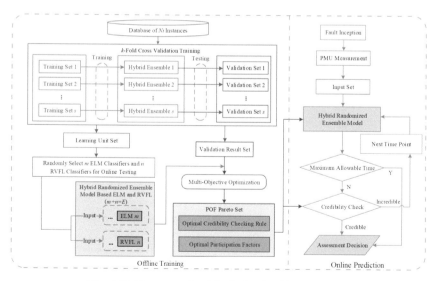

Fig. 6.13: Framework of the hybrid self-adaptive IS

problem. Suppose a database of N_i instances, a k-fold validation process is designed to examine the performance of the hybrid randomized ensemble model. The cross-validation provides a learning unit set and a validation result set. The MOP can be solved to search for the optimal participation rate and credibility checking rule respectively from the two sets. Again, as the previous sections, NSGA-II (K. Deb, *et al.*, 2002) can be applied to solve this MOP, and then a POF can be obtained to interpret the optimization results. In POF, the Pareto points are distributed in a remarkable pattern to show the trade-off relationship between STVS prediction speed and accuracy. Since POF is composed of multiple Pareto points with equal optimality, a compromise solution must be chosen thoughtfully among them to determine the best STVS prediction performance. In general, such compromise Pareto point can be determined by the system operators according to practical requirement. Finally, the trained hybrid randomized ensemble model, the Pareto set, and the POF are available for on-line STVS prediction.

At the on-line application stage, the voltage measurements are assessed self-adaptively over time, using the optimal parameters that are obtained in the off-line training process. The on-line hybrid randomized ensemble model is constructed, based on the optimal participant factors to achieve the highest accuracy, and the credibility of the outputs is evaluated using the optimized credible classification rule to achieve the best overall STVS prediction performance. At each time point, the on-line prediction should also check whether the maximum allowable decision-making time T_{max} has been reached. If T_{max} has been reached, the credibility checking will be skipped, and the SP decision will be directly delivered; if not, only the credible outputs are used as the ultimate STVS prediction decision, and the instances with incredible outputs are pending to next time point for re-assessment.

6.7 Numerical Test for Hybrid Self-adaptive IS

The hybrid self-adaptive IS is tested on the New England 39-bus system and Nordic test system. All the test works are conducted on a computer with a 16 GB RAM and an Intel Core i7 CPU of 3.3 GHz. TDS is performed by using commercial software PSS/E, and the developed IS are implemented in the MATLAB platform.

It is worth mentioning that most of the works in previous sections of the book are tested on the New England 39-bus system, but which may not be large enough to fully demonstrate the strength of the IS methods. Therefore, the numerical test in this section is not only performed on New England system, but also on the Nordic system, which has 23 generators, 41 buses, and 69 branches (*see* Fig. 6.14) to further verify IS effectiveness on larger systems.

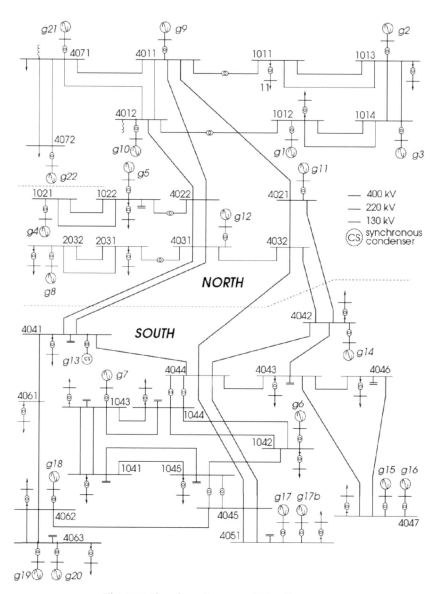

Fig. 6.14: One-line diagram of Nordic system

6.7.1 Database Generation Process

To obtain a comprehensive database for real-time STVS prediction, different physical faults are applied on a wide range of system's operating conditions in order to simulate various post-disturbance system behavior.

In the database generation process, we independently generate 6,000 and 10,000 OPs by randomly varying the load level between 0.8 and 1.2

of its base values for New England 39-bus system and Nordic system, respectively. The generation level from each synchronous generator is determined by OPF. A three-phase fault with a random fault duration between 0.1 and 0.3 seconds and a randomly selected fault location is applied to each OP. Considering the practical situation, the fault is cleared either without loss of power grid component or with a single transmission line tripping, which simulates the different fault-induced topology changes in the database generation. In the end, two separate STVS datasets are generated for the two power systems. The number of stable and unstable instances in each dataset is listed in Table 6.13.

Table 6.13: Number of stable and unstable instances in database

Databases	No. of instances	No. of stable	No. of unstable
New England 39-bus system	6000	3842	2158
Nordic system	10000	5159	4841

6.7.2 Parameter Selection and Tuning

The following parameters need to decided for the hybrid self-adaptive IS:

(1) *Optimal Number of Learning Units in Ensemble*
In the literature, it is shown that when the quantity of single learning units E increases, the error from prediction output gradually decreases and converges to a limit (Z.H. Zhou, *et al.*, 2002; K. Deb, *et al.*, 2002). As shown in Fig. 6.15, the accuracy of ELM and RVFL is stabilized after E = 200. Therefore, in this test, E is selected to be 200.

(2) *Optimal Hidden Node Range and Activation Function*
The number of hidden layer nodes and the choice of activation functions need to be adjusted constantly in ELM and RVFL training process. Under an activation function, the ELM and RVFL classification accuracy can reach the maximum within a particular optimal hidden node range. This is because the learning performance of the different activation functions would deviate in different ways. In this test, only the Sigmoid function is chosen as the valid activation function for both ELM and RVFL classifiers. As shown in Fig. 6.16, the optimal hidden node range of ELM and RVFL algorithms for both New England test system and the Nordic test system is [175, 225].

(3) *Selection of Training and Testing Instances*
The training and testing instances are randomly sampled without replacement from the datasets. In doing so, overlapping between training and testing data is avoided, which means the testing data will be unseen by the trained models. In the meantime, the training and

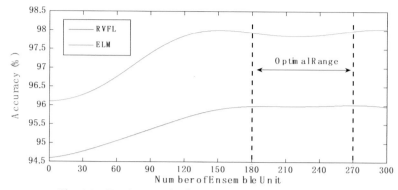

Fig. 6.15: Tuning results for number of learning units

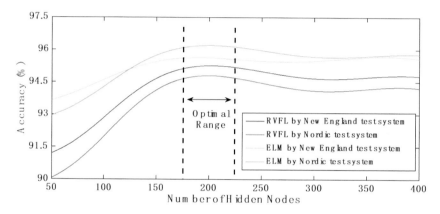

Fig. 6.16: Tuning results for optimal hidden node range

testing instances can both reflect the practical operating circumstances of the power systems. While there is no general criterion for determining the quantity of training and testing instances, based on previous studies and experience, training instance and testing instance are chosen to be 4,800 (80 per cent) and 1,200 (20 per cent) for New England test system, 8,000 (80 pr cent) and 2,000 (20 per cent) for the Nordic test system, respectively.

(4) *Quantity of Training Features*
The decision of how many training features for each time point can affect the whole performance of the robustness and computational efficiency. In this test for New England system and Nordic system, the quantity of features for each time point is controlled at the same level via reducing the sampling frequency.

(5) *Quantity of Cross-validation Folds*
The value of the k decides the quantity of hybrid randomized ensemble models involved in performance validation. The more the hybrid randomized ensemble models are obtained, the better robustness would result. In consideration of the computational efficiency and reliability of performance validation, k is decided to be five.

(6) *Maximum Allowable Decision-making Time T_{max}*
The self-adaptive scheme requires a maximum allowable decision-making time so as to make sure the emergency control can be timely activated. In this case study, T_{max} is defined as 20 s for both New England test system and Nordic test system.

(7) *Time Points*
The New England test system and Nordic test system have different nominal frequencies of 60 Hz and 50 Hz, respectively. In this case, decision cycles of 1/30 s and 1/25 s are used for New England test system and Nordic test system, respectively. A hybrid randomized ensemble model is trained for each post-disturbance time point until T_{max}, so the IS consists of 20 hybrid randomized ensemble models in total.

6.7.3 Performance Validation Optimization Results

To demonstrate the strength of the hybrid randomized ensemble model, the single ensemble models that use ELM and RVFL alone are also tested for comparison. In the test, the credible classification rule is applied to all the tested ensemble models, so the test can purely compare the performances of the hybrid model and the models using a single algorithm.

The MOP is solved by NSGA-II and the corresponding Pareto solutions of three different models are obtained and shown in Fig. 6.17, where the horizontal and vertical axes are respectively average decision speed (ADS) and average decision accuracy (ADA). The key information of the two extreme points in each POF for New England test system and Nordic test system are respectively listed in Tables 6.14 and 6.15.

Table 6.14: Extreme Pareto points for New England test system

Ensemble model	No. of Pareto points	Worst case ADS	Best case ADA	Best case ADS	Worst case ADA
Single ELM	24	3.18	99.89%	1.22	97.33%
Single RVFL	28	2.99	99.54%	1.01	94.23%
Hybrid	28	2.78	99.94%	1.04	97.56%

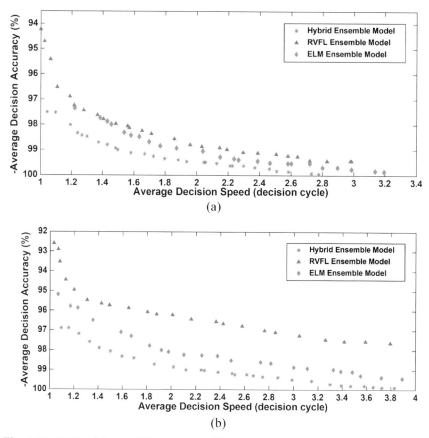

Fig. 6.17: POFs of three different ensemble models on the two test systems: (a) New England 39-bus test system; and (b) Nordic test system

Table 6.15: Extreme Pareto points for Nordic test system

Ensemble model	No. of Pareto points	Worst case ADS	Best case ADA	Best case ADS	Worst case ADA
Single ELM	23	3.86	99.48%	1.07	95.19%
Single RVFL	23	3.79	97.66%	1.03	92.58%
Hybrid	29	3.83	99.91%	1.09	96.90%

From Fig. 6.17, the trade-off relationship between STVS speed and accuracy can be clearly observed, and Pareto points of hybrid randomized ensemble model outperform the other two single ensemble models significantly in terms of accuracy and speed. As Fig. 6.17 shows, the obtained POF from the developed hybrid model is frontier than that of the single models, especially for the Nordic test system. For the extreme Pareto points listed in Tables 6.14 and 6.15, compared to single RVFL ensemble and single ELM ensemble, the hybrid randomized ensemble has the best accuracy for the best case of ADA and the worst case of ADA in New England test system and Nordic test system. For the other extreme Pareto point listed in Tables 6.14 and 6.15, with all the ensemble outputs being identified as credible, the self-adaptive SP of the three models would be instantly completed at around the first time point (i.e. extremely fast SP speed). In this case, the developed hybrid randomized ensemble model can achieve the highest 97.56 per cent and 96.90 per cent ADA, respectively.

In real-time STVS prediction, accuracy and speed are both important. High accuracy serves as a basic requirement in STVS prediction, and a high STVS prediction speed is especially valuable when the system undergoes a fast voltage collapse. In terms of accuracy and speed, the validation result shown in Fig. 6.17 verifies the significance of incorporating multiple learning algorithms rather than adopting a single algorithm.

For computational efficiency, since the hybrid randomized ensemble model adopts the same number (i.e. 200) of learning units as compared to the ensembles of a single learning algorithm, it will not generate additional training burden. The off-line training efficiency of the developed IS and the single ELM and RVFL ensembles are shown in Table 6.16. According to Table 6.16, the off-line training of the hybrid model and the single algorithms consume comparable CPU time, which indicates that the hybrid randomized ensemble of the developed IS is able to improve the STVS prediction performance without impairing its training efficiency.

Table 6.16: Off-line computation efficiency of different models (CPU time)

Ensemble model	Off-line training	
	New England test system	Nordic test system
Single ELM	2532.7 s	3901.2 s
Single RVFL	2563.1 s	3977.3 s
Hybrid	2578.3 s	3935.8 s

The POFs provide a set of optimal solutions, so that the system operators are offered more flexibility in manipulating the SP performance, making the IS more practical in real-world STVS prediction applications. In practice, it is open to system operators to apply their own strategies to select a compromise Pareto point from the POF and satisfy different utility requirements. In this test, the following strategy is used: a minimum

ADA requirement is first established based on the practical utility need to regulate the STVS prediction accuracy and then, the Pareto point with the fastest ADS under the ADA regulation is selected as the compromise solution. For example, suppose the minimum ADA requirement is 99 per cent, then the Pareto points listed in Table 6.17 are shown as the compromise solutions for different ensemble models for New England test system and Nordic test system, since these Pareto points achieve the fastest ADS while satisfying the 99 per cent accuracy requirement. In doing so, the optimal performance of the IS is selected under the accuracy regulation. It can be observed that the hybrid randomized ensemble model is the fastest. Compared with single ELM ensemble and single RVFL ensemble, the hybrid randomized ensemble is 27.5 per cent (18.7 ms) and 35.7 per cent (27.4 ms) faster for New England test system, 37.3 per cent (50.4 ms) (note that single RVFL ensemble cannot achieve the minimum ADA requirement) faster for Nordic test system, respectively. This indicates that the hybrid randomized ensemble model can significantly improve the response time (on average 27.5 per cent - 37.3 per cent faster). In practice, the selection of the compromise Pareto point is not restricted to the strategy adopted in this test. Other decision-making criteria, such as fuzzy logic and Nash equilibrium, can also be applied where appropriate.

Table 6.17: Pareto points of the fastest ADS satisfying the minimum accuracy requirement

Ensemble model	Minimum ADA requirement = 99%	
	New England test system	Nordic test system
Single ELM	68.0 ms	135.2 ms
Single RVFL	76.7 ms	N/A
Hybrid	49.3 ms	84.8 ms

6.7.4 On-line Testing Performance

The hybrid self-adaptive IS is tested by applying the testing instances of New England test system and Nordic test system to show the real-time STVS prediction performance. The selected Pareto points for on-line testing are shown in Table 6.18, and the testing results are listed in Tables 6.19 and 6.20. It can be seen that the ADA and ADS of the New England and Nordic test system in Tables 6.19 and 6.20 are both very close to the validated SP performance in Table 6.18, which indicates the reliability of the POF. More specifically, the ADA in Tables 6.19 and 6.20 are slightly lower than in Table 6.18, but the ADS are faster, which further verifies the trade-off relationship between STVS speed and accuracy.

Table 6.18: Selected Pareto points

New England test system		Nordic test system	
ADS	**ADA**	**ADS**	**ADA**
2.78 decision cycles (92.7 ms)	99.94%	3.83 decision cycles (153.2 ms)	99.91%

Table 6.19: On-line testing performance on New England system

Time points	No. of remaining instances	No. of assessed instances	Current cycle accuracy	Accumulated accuracy
1	1200	569	99.76%	99.76%
2	631	175	100%	99.82%
3	456	105	99.15%	99.73%
4	351	44	97.73%	99.64%
5	307	15	86.67%	99.42%
6	292	25	96.00%	99.34%
7	267	9	100%	99.35%
8	258	13	100%	99.35%
9	245	0	N/A	99.35%
10	245	6	100%	99.36%
11	239	26	96.15%	99.27%
12	213	3	100%	99.28%
13	210	7	100%	99.28%
14	203	6	100%	99.29%
15	197	1	100%	99.29%
16	196	12	91.67%	99.14%
17	184	2	100%	99.15%
18	182	3	100%	99.15%
19	179	7	71.43%	99.07%
20	172	1	100%	99.07%
ADS	2.69 decision cycles (89.7 ms)		ADA	99.36%
Time-delay	50 ms			
Total response time	139.7 ms			

Table 6.20: On-line testing performance on Nordic system

Time points	No. of remaining instances	No. of assessed instances	Current cycle accuracy	Accumulated accuracy
1	2000	703	99.86%	99.86%
2	1297	121	100%	99.88%
3	1176	407	99.26%	99.68%
4	769	75	97.33%	99.54%
5	694	25	88.00%	99.32%
6	669	95	96.84%	99.16%
7	574	14	100%	99.17%
8	560	23	100%	99.18%
9	537	2	100%	99.18%
10	535	10	100%	99.19%
11	525	43	95.35%	99.08%
12	482	6	100%	99.08%
13	476	11	100%	99.09%
14	465	10	100%	99.10%
15	455	2	100%	99.10%
16	453	35	94.29%	98.99%
17	418	4	100%	98.99%
18	414	13	100%	99.00%
19	401	12	75.00%	98.82%
20	389	20	100%	98.85%
ADS	3.56 decision cycles (142.4 ms)			
Time-delay	50 ms		ADA	99.21%
Total response time	192.4 ms			

According to IEEE Standard C37.118.2-2011 (I.S. Association, 2011), the typical communication delay between PMU and PDC is 20 ms to 50 ms. Such time-delay in PMU communication depends on the data acquisition infrastructure and will inevitable postpone the STVS prediction response. To take such impact into account, a flat time-delay of 50 ms is assumed and added on the response time. As a result, the total response time is 139.7 ms and 192.4 ms, respectively, as shown in Tables 6.19 and 6.20. Although the STVS prediction is further delayed by 50 ms, the developed IS still completes the SP at the earliest opportunity to allow the most time for emergency control. The test results also indicate that as the power system becomes more complicated (e.g. change from New England to Nordic system), the IS tends to acquire more data to achieve the same level of accuracy, which postpones the STVS prediction (e.g. 49.3 ms for New England system vs 84.8 ms for Nordic system as shown in Table 6.17). In this situation, the SP speed improvement by the developed IS becomes increasingly significant for larger systems (e.g. 18.7 ms for New England system vs 50.4 ms for Nordic system as indicated in Table 6.17). Considering more complicated power systems in practice, the impact of communication time-delay tends to diminish while the SP speed merit of the hybrid self-adaptive IS is more clearly demonstrated.

6.7.5 Guidance on Voltage Stability Control

For practical application, the earlier SP result provided by the developed IS can trigger faster and greater timely stability controls than traditional response-based control, such as UVLS. Owing to its higher SP speed, the self-adaptive IS can accurately predict the voltage stability state at an earlier stage. This triggers the emergency control actions (e.g. load shedding) earlier when an unstable voltage propagation is anticipated. Such early activation of emergency control, not only contributes to improving the chance of regaining voltage stability, but also reduces the load shedding amount and economic loss. At a longer response time, the system tends to be closer to its voltage instability boundary and a larger control force would be needed to drag the system back to its stable equilibrium, resulting in a higher amount of load shed.

To illustrate this, a load shedding test is performed. In this test, an unstable New England system instance is arbitrarily selected from the dataset. Load shedding at different response times and UVLS are tested on the selected instance to compare their control effectiveness and economic efficiency. Three response times are selected for load shedding – 0.14 s (i.e. the response time of the IS plus time-delay), 0.4 s (i.e. a later response time plus time-delay), and 0.72 s (i.e. the maximum allowable decision-making time plus time-delay). The UVLS scheme adopted in the test is shown in Table 6.21 where the voltage threshold and time-delay settings are adjusted to reduce the total load shedding amount. The voltage stability state and the load shedding percentages are given in Table 6.22, and the voltage trajectories with and without the control actions are shown in Fig. 6.18.

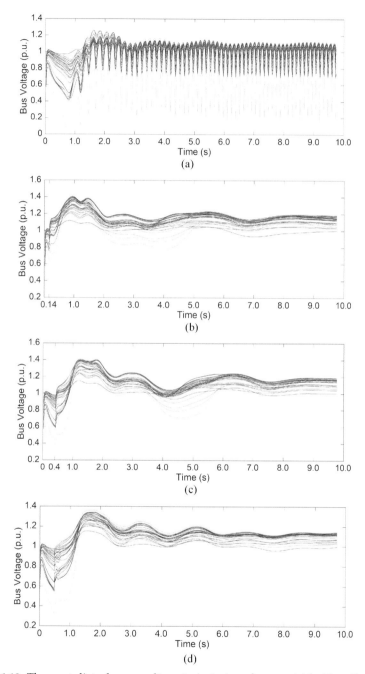

Fig. 6.18: The post-disturbance voltage trajectories of an unstable New England system instance in different control situations: (a) without load shedding, (b) with load shedding at 0.14 s, (c) with load shedding at 0.4 s, (d) with UVLS

Table 6.21: The UVLS scheme

Step	Voltage threshold (p.u.)	Time delay (s)	Load shedding percentage (%)
1	0.9	0.4	20
2	0.85	0.4	20
3	0.8	0.4	20

Table 6.22: Load shedding effectiveness and amount

	Load shedding based on the proposed method			UVLS
	0.14 s	0.4 s	0.72 s	
Voltage stability state after control	Stable	Stable	Unstable	Stable
Load shedding percentage	8.9 %	15.8 %	N/A	29.6%

It can be observed that, if the control action is activated too late (e.g. 0.72 s where the voltage has already collapsed), the control would fail to regain voltage stability. By contrast, the earlier the load is shed, the less load-shedding amount is needed (comparing 0.14 s and 0.4 s). The results also show that the UVLS needs more load-shedding amount because the load is shed based on the voltage magnitude and there is no voltage SP. For the developed hybrid self-adaptive IS, the load can be immediately shed at 0.14 s when the method delivers a reliable voltage SP result as 'unstable'. Hence, it has least load shedding amount and earlier voltage-stability recovery.

6.8 Probabilistic Self-adaptive IS for Post-disturbance STVS Prediction

In previous Sections 6.4 and 6.6, the self-adaptive STVS prediction was implemented via ensemble learning, output aggregation, and MOP optimization. The ensemble models predict the voltage stability status/ FIDVR severity at different time points. The output aggregation determines the credibility of the ensemble outputs under a rule-of-thumb scheme. The MOP optimizes the parameters involved in the output aggregation to achieve the best STVS prediction performance in terms of accuracy and speed.

However, such self-adaptive process still shows the following drawbacks:

- The rule-of-thumb scheme for output aggregation lacks mathematical rigorousness.

- A large number of user-defined parameters are involved, and the optimization on those parameters is time-consuming.
- The parameters are optimized at the aggregation performance validation stage. Such off-line validated performance cannot reliably reflect the on-line STVS prediction performance.

Considering the above drawbacks, this section aims at more sophisticated implementation of self-adaptive STVS prediction, and develops a probabilistic self-adaptive IS for real-time STVS prediction. The IS is separately designed for voltage instability detection and FIDVR prediction. The main features, of the developed IS, are as follows:

- Self-adaptive detection of voltage-instability events uses probabilistic classification models. Following a disturbance, the probability of the voltage being unstable is monitored by the probabilistic classification models over a progressively increasing time window, and the emergency actions are triggered once insufficiently high probability is detected. In doing so, the unstable events are detected in a self-adaptive way, which enhances the overall voltage instability detection speed without impairing the detection accuracy.
- Self-adaptive SP on FIDVR events use probabilistic prediction models. At each sliding time window following a disturbance, the probabilistic prediction models not only provide a deterministic prediction on FIDVR severity, but also model the error existing in the prediction process. In doing so, the FIDVR severity is predicted on a probabilistic basis with a certain confidence level. The SP decision is delivered when sufficient decision confidence is reached, so the SP result can be obtained earlier without sacrificing the accuracy.
- The probabilistic self-adaptive IS is non-parametric in nature (no user-defined parameter is involved), which avoids high parameter-tuning burden and improves the reliability and robustness of STVS prediction under the self-adaptive architecture.

6.8.1 RVFL Ensemble Model

Another emerging learning algorithm, RVFL, is adopted to build the probabilistic self-adaptive IS for real-time FIDVR prediction. Similar to ELM, RVFL is a RLA and has shown significant advantages in ensemble learning. As such, the probabilistic predictors of the developed IS are designed on the basis of an RVFL ensemble model.

In the RVFL ensemble model, the training data to each individual RVFL is sampled, using pair bootstrap method which is a general statistical inference approach for regression analysis (A. Khamis, *et al.*, 2016). Then the RVFL ensemble model is constructed as follows:

(1) Generate N bootstrapped pairs $\{(\mathbf{x}_1^*, \mathbf{t}_1^*), ..., (\mathbf{x}_N^*, \mathbf{t}_N^*)\}$ by uniform sampling with replacement from the original database $\{(\mathbf{x}_1, \mathbf{t}_1), ..., (\mathbf{x}_N, \mathbf{t}_N)\}$.

(2) Randomly assign the number of hidden nodes of RVFL from the optimal range $[h_{min}, h_{max}]$.

(3) Train a RVFL using the N bootstrapped pairs.

(4) Repeat (2) and (3) until N_R RVFLs is obtained.

(5) The N_R RVFLs constitute the RVFL ensemble model.

The RVFL, as an RLA, is selected as the single learning unit to increase diversity in the ensemble model from three aspects: data diversity, structure diversity, and parameter diversity. How the three diversities are achieved in the learning process of RVFL ensemble model is illustrated in Fig. 6.19.

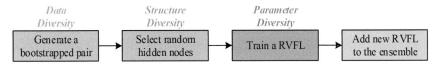

Fig. 6.19: Learning diversity achieved in the RVFL ensemble learning process

6.8.2 Probabilistic Self-adaptive Voltage Instability Detection

In real-time STVS prediction, the unstable voltage must be detected at an early stage to activate an effective emergency control action. This means that the SP is actually performed on a forecasting basis. In the literature, the existing methods (S Dasgupta, *et al.*, 2013; L. Zhu, *et al.*, 2015; L. Zhu, *et al.*, 2017) can only provide deterministic SP results, i.e. classify the post-disturbance voltage propagation into either a stable or an unstable event. Such deterministic results cannot be fully accurate due to the forecasting uncertainty. To mitigate this issue, this section presents a probabilistic self-adaptive IS for real-time voltage instability detection, aiming to improving the detection speed without impairing the detection accuracy through a non-parametric approach.

6.8.2.1 The Probabilistic Self-adaptive Implementation

In the literature, the existing methods for real-time STVS prediction adopt a long fixed-length time window following the disturbance. Due to the forecasting nature, real-time STVS prediction is exposed to higher forecasting uncertainty for shorter observation. So the time window must be sufficiently long to achieve acceptable SP accuracy, which impairs the SP speed. In contrast, the probabilistic self-adaptive IS can monitor the probability of voltage instability over a progressively increasing time window, and detect the unstable events once the probability reaches the limit. In doing so, the unstable event can be detected as early as possible, which improves the overall SP speed.

The probabilistic self-adaptive implementation for voltage instability detection is illustrated in Fig. 6.20. After the fault is cleared at T_c, the post-disturbance STVS prediction is performed over a number of time points. At each time point T_k, $k \in 1, 2, ..., F$, probabilistic classification is performed based on the voltage trajectories collected within the time-window between T_c and T_k, and the detection decision is made by simply comparing the probability of voltage instability p with a warning limit p_{max}. If p is over the limit, an unstable SP decision will be issued and the corresponding emergency control actions will be activated. On the other hand, if p is below the limit, the evidence to support an unstable SP decision is considered insufficient under the current time window. Then the voltage is re-assessed under a longer time-window. Such a self-adaptive process continues until an unstable SP decision is made or the maximum decision time is reached. In practice, a maximum decision time T_F is needed to define the latest time when the SP decision must be delivered. At T_F, p is compared with 50 per cent to decide the stability status. A p greater than 50 per cent means that the voltage is more likely to be unstable. So an unstable SP decision is delivered. If p is smaller than 50 per cent, the voltage is assessed as stable.

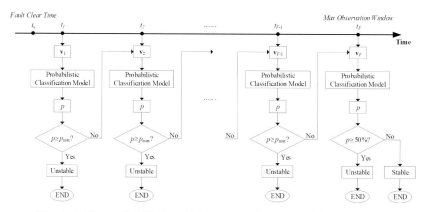

Fig. 6.20: The probabilistic self-adaptive voltage instability detection

In Sections 6.4 and 6.6, the self-adaptiveness of voltage instability detection was achieved via a rule-of-thumb scheme that involves a number of user-defined parameters. In practice, the SP performance in terms of accuracy and speed is sensitive to the values of those parameters, meaning time-consuming parameter tuning process is a must. In contrast, the probabilistic self-adaptive IS captures the inherent probability of the different voltage stability status, which is non-parametric in nature. Such a non-parametric method is easier to be implemented, and is more reliable and more robust in real-world applications.

6.8.2.2 Probabilistic Classification Model

For a classification task, a successful probabilistic classification model should be able to numerically describe the uncertainties existing in the problem, and estimate the probability of each class. For voltage instability detection, the result from a probabilistic classification model is the probability of the post-disturbance voltage being stable or unstable. To achieve this, the probabilistic classification model is constructed, based on the framework in (C. Wan, *et al.*, 2014). Its working principle is shown in Fig. 6.21.

The probabilistic classification model consists of two ensemble models. At off-line stage, the two ensemble models are sequentially trained as follows: an ensemble model A is trained as a regression model using the database $\mathcal{D} = \{(\mathbf{x}_i, t_i)\}_{i=1}^{N}$. Then an ensemble model B is trained as a regression model using a modified database $\mathcal{D}_e = \{(\mathbf{x}_i, e_i)\}_{i=1}^{N}$. Such database modification is to replace the output of each instance by the training error of ensemble model A. The single learning units in the two ensemble models are respectively denoted as $\{h_1^A, \dots, h_S^A\}$ and $\{h_1^B, \dots, h_S^B\}$, where S is the number of single learning units in each ensemble model. Then the training error e_i is

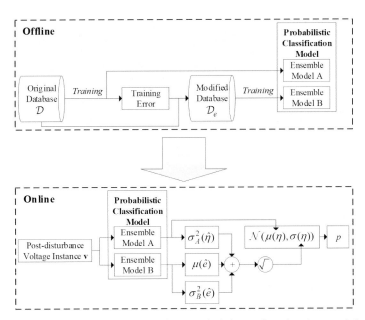

Fig. 6.21: Working principle of the probabilistic classification model

$$e_i = \left(t_i - \frac{1}{S}\sum\nolimits_{j=1}^{S} h_j^A(\mathbf{x}_i) \right)^2 \tag{6.18}$$

At on-line stage, for an incoming post-disturbance voltage instance $\mathbf{v}$, ensemble model A is responsible to predict its stability margin η, and the ensemble model B is responsible to estimate the prediction uncertainty. Here, the stability margin η is a numerical quantity to describe the trend of the voltage to propagate into stable or unstable status. For example, if we respectively label stable and unstable status as 1 and –1, $\eta = 0.1$ indicates that the voltage has a weak trend to be stable, while $\eta = -0.9$ indicates a very strong unstable trend. Such stability margin concept suits the needs of real-time STVS prediction with forecasting uncertainties. It facilitates probabilistic classification. Based on the outputs from the two ensemble models, distribution of η is estimated as follows:

$$\eta \sim \mathcal{N}(\mu(\eta), \sigma(\eta)) \tag{6.19}$$

where

$$\mu(\eta) = \frac{1}{S}\sum\nolimits_{j=1}^{S} h_j^A(\mathbf{v}) \tag{6.20}$$

$$\sigma(\eta) = \sqrt{\sigma_A^2(\hat{\eta}) + \mu(\hat{e}) + \sigma_B^2(\hat{e})} \tag{6.21}$$

In (6.19)-(6.21), Gaussian distribution is assumed for η. $\mu(\eta)$ and $\sigma^2(\eta)$ respectively represent the mean and variance of the distribution. The mean is estimated as the average output from ensemble model A, and the variance is formulated as the sum of three components to fully capture the uncertainties existing in the prediction process. The three components are the average ensemble model B output, $\mu(\hat{e})$, to evaluate the expected prediction variance; the ensemble model A output variance, $\sigma_A^2(\hat{\eta})$, to evaluate the inherent model uncertainty of ensemble model A, and the ensemble model B output variance, $\sigma_B^2(\hat{e})$, to evaluate the inherent model uncertainty of ensemble model B. These three components can be mathematically calculated as follows:

$$\mu(\hat{e}) = \frac{1}{S}\sum\nolimits_{j=1}^{S} h_j^B(\mathbf{v}) \tag{6.22}$$

$$\sigma_A^2(\hat{\eta}) = \frac{1}{S-1}\sum\nolimits_{j=1}^{S} (h_j^A(\mathbf{v}) - \mu(\eta))^2 \tag{6.23}$$

$$\sigma_B^2(\hat{e}) = \frac{1}{S-1}\sum\nolimits_{j=1}^{S} (h_j^B(\mathbf{v}) - \mu(\hat{e}))^2 \tag{6.24}$$

Based on the Gaussian distribution, the probability of an unstable event p is estimated as follows:

$$p = P(\eta \leq 0) \tag{6.25}$$

where $P(\cdot)$ is a function to calculate the probability of the condition specified in the bracket. By using –1 and 1 to label unstable and stable status, $\eta \leq 0$ indicates an unstable voltage propagation following the disturbance.

6.8.3 Probabilistic Self-adaptive FIDVR Prediction

The FIDVR generally is assessed via a time-series decision-making process to decide whether a FIDVR event is acceptable or not. This section presents the probabilistic self-adaptive IS for FIDVR prediction. This IS is designed to improve the FIDVR prediction speed without impairing the prediction accuracy through a non-parametric approach. In this section, TVSI is employed to evaluate the severity of FIDVR events.

6.8.3.1 The Probabilistic Self-adaptive Implementation

In the related literature (S.M. Halpin, *et al.*, 2008; H. Bani and V. Ajjarapu, 2010; Y. Dong, *et al.*, 2017; L. Zhu, *et al.*, 2015; L. Zhu, *et al.*, 2017), the data-driven methods use a pre-selected input vector (e.g. a certain length of system trajectories) to classify/predict the system's stability status. To decide if a FIDVR event is acceptable, such a method, if adopted, will directly predict the TVSI value based on the real-time voltage trajectories, and make the final decision by simply comparing the predicted TVSI value with the predefined TVSI threshold. If the TVSI value is predicted to be higher than the TVSI threshold, the FIDVR event is assessed as unacceptable, and subsequent emergency control actions, such as load shedding, are triggered. However, the error existing in such TVSI prediction process cannot be properly captured and controlled, so a greater prediction accuracy is only achievable at the cost of later decision time when more voltage trajectory data becomes available. In most cases, the SP decision-making speed may be too slow to timely trigger emergency control actions.

To improve the time-series decision-making speed, the IS can be implemented in a probabilistic self-adaptive way. It progressively predicts the TVSI based on the real-time voltage snapshot data collected within each sliding time window. The principle is that the TVSI is predicted on a probabilistic basis with a certain confidence level, and the final FIDVR prediction decision is made once sufficient decision confidence is achieved. Using this IS, the FIDVR prediction decision can be reliably delivered as early as possible without impairing the accuracy. Thus overall decision-making speed will be significantly improved.

The probabilistic self-adaptive IS for real-time FIDVR prediction is illustrated in Fig. 6.22 where the curves at the top represent the post-disturbance voltage trajectories collected in real-time; T_c is the fault

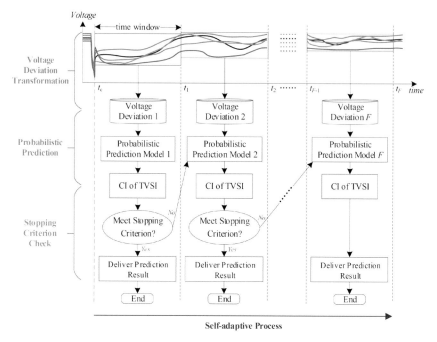

Fig. 6.22: The probabilistic self-adaptive FIDVR prediction process

clearing time; $T_1...T_F$ are the time points when the FIDVR prediction action is progressively executed, and T_F is the latest decision time after which the emergency control actions must be taken to be effective. Based on the F time points, F sliding time windows are defined as the time period between each of the two consecutive time points.

The probabilistic self-adaptive process is implemented as follows: at each time point, a snapshot of voltage trajectories is collected within the time window between current and the previous time points, and the voltage magnitude is transformed into voltage deviation. Based on the trajectories of voltage deviation, TVSI is probabilistically predicted by a probabilistic prediction model. The prediction result comes in the form of the confidence interval (CI) which is the most probable range where the true TVSI value will fall with a certain confidence level. In doing so, the probabilistic prediction not only predicts the deterministic TVSI value, but also models the error existing in the prediction process. Based on the predicted CI, a stopping criterion is checked to decide whether a FIDVR prediction decision can be made with sufficient confidence. If the stopping criterion is not met, the probabilistic prediction and stopping criterion check progress at the next time point with the updated voltage trajectory snapshot. In practice, a maximum decision time T_F is needed, at which time the FIDVR prediction decision must be delivered regardless of the stopping criterion. Since F time windows are defined, F probabilistic predictors should be prepared respectively for different time windows.

6.8.3.2 Voltage Deviation Transformation

Different from the probabilistic classification for voltage instability detection, each probabilistic prediction for FIDVR prediction is performed based on the voltage trajectory snapshot between the current and the previous time points. In doing so, the input data size to each probabilistic predictor is dramatically reduced, so that the computational efficiency of probabilistic prediction can be significantly improved. The time window width ΔT is determined based on the effectiveness of the voltage trajectory snapshot and the computation time of each prediction.

By definition, TVSI is computed based on the voltage deviation level, which is the ratio between the deviated voltage and the pre-disturbance voltage magnitude. Therefore, rather than voltage magnitude, using voltage deviation as the input features in the probabilistic prediction models helps improve the prediction accuracy on TVSI. The voltage deviation on bus i at time t, $VD_{i,t}$, is computed as:

$$VD_{i,t} = \frac{\left|V_{i,t} - V_{i,0}\right|}{V_{i,0}} \tag{6.26}$$

6.8.3.3 Confidence Interval Design

Probabilistic prediction not only deterministically predicts the TVSI value, but also models the uncertainty existing in data, models, and prediction process. The results of probabilistic prediction come in the form of the CI. In the IS, it is assumed that the deterministic prediction result follows a Gaussian distribution, so the typical formulation of a CI with $100(1 - \alpha)$ per cent confidence level is

$$CI^{\alpha}(t_i) = [L^{\alpha}(t_i), U^{\alpha}(t_i)] \tag{6.27}$$

$$L^{\alpha}(t_i) = \hat{t}_i - z_{1-\alpha/2}\sqrt{\sigma^2(t_i)} \tag{6.28}$$

$$U^{\alpha}(t_i) = \hat{t}_i + z_{1-\alpha/2}\sqrt{\sigma^2(t_i)} \tag{6.29}$$

where CI is an interval enclosed by a lower bound $L^{\alpha}(t_i)$ and an upper bound $U^{\alpha}(t_i)$; $z_{1-\alpha/2}$ is the critical z-value of the standard Gaussian distribution; $\hat{t}_i$ denotes the predicted value of t_i; and $\sigma^2(t_i)$ is the prediction variance on t_i, thereby representing the error in the prediction process.

In the probabilistic self-adaptive IS for FIDVR prediction, the prediction target t_i is the TVSI value of the ith FIDVR instance, and a CI with $100(1 - \alpha)$ per cent confidence level means that the probability of the true TVSI value within the CI is $100(1 - \alpha)$ per cent. By definition, the computed TVSI for a FIDVR event should always be positive; so any negative $L^\alpha(t_i)$ will be adjusted to zero.

6.8.3.4　Stopping Criterion

In the self-adaptive process, the performance of probabilistic prediction determines the quality of the CI, and the stopping criterion regulates the accuracy of the delivered SP decision at each time point before T_F. Therefore, the design of the stopping criterion is crucial for the self-adaptive performance. Based on the predicted CI, the self-adaptive process stops subject to the following criterion:

$$\textbf{If}\begin{cases} L^\alpha(t_i) \geq \overline{\text{TVSI}} \Rightarrow \text{Unacceptable FIDVR event} \\ U^\alpha(t_i) \leq \overline{\text{TVSI}} \Rightarrow \text{Acceptable FIDVR event} \end{cases} \tag{6.30}$$

where $\overline{\text{TVSI}}$ is the practical TVSI threshold to distinguish acceptable and unacceptable FIDVR events. If the true TVSI value is beyond $\overline{\text{TVSI}}$, the FIDVR event is then considered as unacceptable, and vice versa. The value of $\overline{\text{TVSI}}$ is empirically selected based on practical requirements on FIDVR. In the stopping criterion (6.30), the predicted CI is compared with the TVSI threshold to make the final SP decision. In doing so, reliable decisions are made, based on the probabilistic prediction result. Since a certain level of TVSI prediction confidence is regulated by the predicted CI and the SP decision is made only when the CI fully locates at one side to the TVSI threshold, the prediction accuracy at each time point should be higher than the pre-defined confidence level. For instance, for a predicted CI with 99 per cent confidence level, the probability of the true prediction target t_i falling out of the predicted CI is 1 per cent. In this case, there are two possible scenarios – either $t_i < L^\alpha(t_i)$ or $t_i < U^\alpha(t_i)$. As Gaussian distribution is assumed in constructing the CI, the probability of those two scenarios should be equally 0.5 per cent. Note that only one of those two scenarios can actually result in misclassification on the FIDVR event, so at least 99.5 per cent FIDVR prediction accuracy can be achieved by the predicted CI at each time point.

For a specific FIDVR event, its SP decision tends to be more accurate at a later time point because more voltage trajectory data becomes available. In the CI definition, such higher accuracy can be reflected as a smaller error term $\sigma^2(t_i)$. Thus the predicted CI tends to be narrower at a later time point. Then, according to (6.30), the probability of the stopping criterion to be met should be higher for narrower CI, so that there is a higher probability for the FIDVR event to be successfully assessed at a later time point; meaning, the self-adaptive process is achievable and reasonable. Moreover, the stopping criterion is parameter-free, which is a significant advantage over the self-adaptive method in Sections 6.4 and 6.6 that involve a number of user-defined parameters.

In probabilistic self-adaptive FIDVR prediction, the stopping criterion (6.30) is applied only up to the time point T_{F-1}. If the SP result is not delivered until the latest decision time T_F, the SP decision is compulsorily made, based on the value of $\hat{t}_i$. In this case, if $\hat{t}_i$ is larger than $\overline{TVSI}$, the FIDVR event is assessed as unacceptable, and vice versa.

6.8.3.5 *Probabilistic Prediction Model*

In the developed IS, the predicted CI at each time window determines the overall decision-making accuracy and speed. Therefore, the probabilistic prediction model to predict each CI should be carefully designed. Its structure is similar to the probabilistic prediction model in Section 6.8.2.2 which consists of two ensemble models. One of them, called 'TVSI predictor', works for deterministically predicting the TVSI values based on the incoming trajectory snapshot, and the other one, called 'error predictor', works for estimating the error between the fitted and the true TVSIs. RVFL ensemble model is employed for both TVSI predictor and error predictor. The principle of the probabilistic prediction model is to use the multiple and diversified RVFL outputs to statistically estimate the Gaussian distribution of the prediction results, and the CI can be constructed using (6.27) - (6.29) based on the estimated Gaussian distribution. The working principle of the probabilistic prediction model is illustrated in Fig. 6.23.

At the off-line stage, the TVSI predictor and the error predictor are sequentially trained. For each sliding time window, a TVSI predictor, an ensemble of N_R RVFLs, is firstly trained by the TVSI training set. Such a training set consists of the trajectory snapshot within the time window as the input features and the computed TVSI value as the known target for each FIDVR instance. The associated training errors in RVFL ensemble training are extracted as follows:

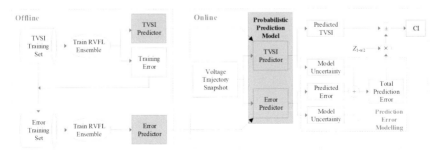

Fig. 6.23: Working principle of the probabilistic prediction model

$$e_j = \left(t_j - \frac{1}{N_R} \sum_{r=1}^{N_R} y_{\text{fit},r}(t_j) \right)^2 \qquad (6.31)$$

where e_j and t_j are the training error and the true TVSI value of training instance j, and $y_{\text{fit},r}(t_j)$ denotes the fitted TVSI value of training instance j by the rth RVFL in the TVSI predictor. These training errors then replace the TVSI values as the training targets, and the original TVSI training set is transformed into an error training set. The error predictor, another ensemble of N_R RVFLs, is then trained by such error training set to model the fitting error of the TVSI predictor.

At the on-line stage, if a fault is cleared in the system, the F probabilistic prediction models take their turns to predict CI. The CI is predicted based on the trajectory snapshot collected at each time point. Although the predicted CI should enclose both the predicted TVSI and the prediction error, these two parts are computed separately at the on-line stage.

The predicted TVSI is the mean output from individual RVFLs in the TVSI predictor

$$\hat{t}_i = \frac{1}{N_R} \sum_{r=1}^{N_R} y_r(t_i) \qquad (6.32)$$

where $\hat{t}_i$ is the predicted TVSI value and $y_r(t_i)$ is the output on instance i from the rth RVFL in the TVSI predictor.

The prediction error in CI is modeled by not only the expected error between the fitted and the true target values, but also the inherent uncertainty of the intelligent models (C. Wan, et al., 2014). Therefore, the modeling of the prediction error $\sigma^2(t_i)$ consists of three components: the model uncertainty from the TVSI predictor, the expected error, and the model uncertainty from the error predictor. They are mathematically formulated as follows:

$$\sigma^2(t_i) = \sigma_M^2(t_i) + \hat{e}_i + \sigma_{ME}^2(t_i) \qquad (6.33)$$

where

$$\sigma_M^2(t_i) = \frac{1}{N_R - 1} \sum_{r=1}^{N_R} (y_r(t_i) - \hat{t}_i)^2 \qquad (6.34)$$

$$\hat{e}_i = \frac{1}{N_R} \sum_{r=1}^{N_R} y_{e,r}(t_i) \qquad (6.35)$$

$$\sigma_{ME}^2(t_i) = \frac{1}{N_R - 1} \sum_{r=1}^{N_R} (y_{e,r}(t_i) - \hat{e}_i)^2 \qquad (6.36)$$

where $\sigma_M^2(t_i)$ and $\sigma_{ME}^2(t_i)$ denote the model uncertainty from the TVSI predictor and the error predictor respectively, $\hat{e}_i$ is the expected error of t_i,

$y_{e,r}(t_i)$ is the output of instance i from the rth RVFL in the error predictor. The predicted CI based on such prediction error modeling is highly reliable, which has been verified in (C. Wan, *et al.*, 2014) using the low average coverage error index and the interval score index.

6.9 Numerical Test for Probabilistic Self-adaptive IS

The numerical tests for voltage instability detection and FIDVR prediction are performed separately to verify the performance of the probabilistic self-adaptive IS. Section 6.9.1 presents the voltage instability detection test, whereas the FIDVR prediction test is elaborated in Section 6.9.2. Although voltage instability detection and FIDVR prediction are tested separately, they can still be coordinated under the hierarchical architecture presented in Section 6.4.

Both the tests are carried on the database created in Section 6.5.1. In the voltage instability detection test, the original database of 9,727 instances is used. Among them, 2,000 instances are randomly selected for testing purpose, and the remaining instances are for training purpose. In the FIDVR prediction test, only the stable instances in the original database are selected to construct the FIDVR database. As a consequence, the FIDVR database consists of 5,821 instances. Among them, 4,657 (80 per cent) instances are used as the training instances, and the rest 1,164 (20 per cent) are used to test the FIDVR prediction performance.

6.9.1 Voltage Instability Detection Test

In the voltage instability detection test, ELM is employed as the machine learning algorithm for the two ensemble models to construct the probabilistic classification models. The overall SP performance of the IS is evaluated in terms of accuracy and speed. The SP performance at each time window is also investigated. In the end, the probabilistic self-adaptive IS is compared with Lyapunov exponent method (S. Dasgupta, *et al.*,2013) to verify its superiority.

6.9.1.1 Setup for Self-adaptive Stability Prediction

The following quantities should be decided to implement the self-adaptive process for voltage instability detection.

- *Maximum observation window, T_F*
  The maximum observation window T_F decides the latest time to detect unstable events. Longer T_F improves the SP accuracy, but may result in ineffective emergency control. In the test, a T_F of 1 second is adopted.

- *Time windows, T_1 to T_{F-1}*
  In a self-adaptive process, the setting of time windows decides the SP resolution. If the resolution is higher, more probabilistic classification models need to be trained. In the test, the probabilistic classification is performed every 0.1 second, so that the time window progressively increases from 0.1 second to 1 second with a step size of 0.1 second. With such setting, 10 probabilistic classification models are trained.
- *Warning limit, p_{max}*
  The warning limit p_{max} is the threshold to detect unstable events and trigger emergency control actions, so that it is sufficiently high to avoid a false alarm. In the test, p_{max} is 99 per cent.
- *Number of single learning units in ensemble model, S*
  A larger S can improve the probabilistic classification performance, but increases the off-line training burden. In the test, S is set at 200.

6.9.1.2 Performance Metrics

The main contribution of self-adaptive STVS prediction is to improve the voltage instability detection speed without impairing the accuracy. Thus appropriate metrics should be defined to evaluate its performance. Similar to the TSP test in Section 6.3, the overall voltage instability detection performance in this test is evaluated using two metrics – mean assessment accuracy (MAA), and mean detection time (MDT). They are mathematically calculated as follows:

$$MAA = \frac{1}{S+U}(\sum_{k=1}^{F-1} A_k N_k + A_F(S+U-\sum_{k=1}^{F-1}N_k)) \qquad (6.37)$$

$$MDT = \frac{1}{U}(\sum_{k=1}^{F-1} t_k A_k N_k + t_F(U-\sum_{k=1}^{F-1}A_k N_k)) \qquad (6.38)$$

where S and U are respectively the numbers of stable and unstable testing instances; N_k ($k < F$) are the numbers of detected instances at each time window before T_F; and A_k ($k \leq F$) are the prediction accuracy at each time window. In the self-adaptive process, only unstable events can be detected earlier, but the SP on stable events is left to the maximum time window. So the calculation of MDT only accounts for the unstable instances to evaluate the early voltage instability detection ability of the developed IS.

6.9.1.3 Testing Results

The voltage instability detection test proceeds in a self-adaptive way over a progressively increasing time window. Under such self-adaptive architecture, the testing performance at each time window, together with the overall testing performance, is shown in Table 6.23. It should be noted that the time windows between 0.1 second and 0.9 second are for detection of unstable instances, and the 1.0 second time window is for

Table 6.23: Testing results of probabilistic self-adaptive voltage instability detection

Time windows	No. of remaining instances	No. of assessed instances	Accuracy (i.e. A_k)
0.1 s	2000	938	100%
0.2 s	1062	5	100%
0.3 s	1057	4	100%
0.4 s	1053	11	100%
0.5 s	1042	7	100%
0.6 s	1035	5	100%
0.7 s	1030	4	100%
0.8 s	1026	3	100%
0.9 s	1023	2	100%
1.0 s (t_F)	1021	1021	98.14%
MDT	0.139 s	**MAA**	99.05%

the SP of all remaining instances, including both the undetected unstable instances and the stable instances. In such a self-adaptive process, 100 per cent accuracy at 0.1 second – 0.9 second time windows means that the earlier detection of unstable events does not generate any miss alarm. This verifies the ability of the developed IS to improve the detection speed without impairing the detection accuracy.

6.9.1.4 Comparison with Lyapunov Exponent Method

The probabilistic self-adaptive IS for voltage instability detection is compared with the Lyapunov exponent method in (S. Dasgupta, *et al.*, 2013) to further demonstrate its fast SP ability. The overall testing performance of the Lyapunov exponent method is shown in Table 6.24. Compared to the developed IS, the Lyapunov exponent method results in a similar level of MAA (99.05 per cent for the developed IS vs 99.00 per cent for the Lyapunov exponent method) with much longer MDT (0.139 second for the developed IS vs 0.398 second for the Lyapunov exponent method). Such a result verifies that the developed IS significantly increases the voltage instability detection speed without impairing the accuracy. Moreover, the detection time of each unstable instance using the two methods is shown in Fig. 6.24. Here, the detection time of unstable instances using Lyapunov exponent method is defined as the time when the Lyapunov exponent series exceeds zero. In Fig. 6.24, it is observed that the Lyapunov exponent method leads to much longer detection time than the developed IS for almost all the unstable instances.

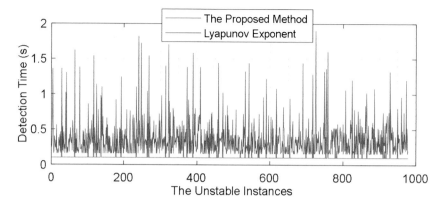

Fig. 6.24: Comparison of instability detection time between the probabilistic self-adaptive IS and the Lyapunov exponent method

Table 6.24: Overall testing performance of Lyapunov exponent method

MAA	MDT
99.00%	0.398 s

6.9.2 FIDVR Prediction Test

The FIDVR prediction test only considers the stable instances from the TDS, and the true TVSI values of these stable voltage trajectories are calculated using (1.10)-(1.11) as the prediction target. The distribution of the TVSI value is shown in Fig. 6.25 where the TVSI value of most instances lie between 0 and 6.

Besides testing the accuracy and speed, more comprehensive evaluations on the probabilistic self-adaptive IS are conducted in this test, including the reliability of the predicted CI, the computational efficiency of

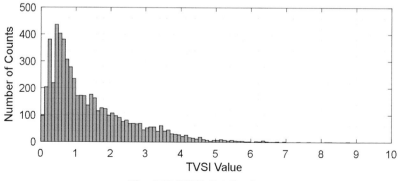

Fig. 6.25: TVSI distribution

the probabilistic prediction model, sensitivity analysis, and comparative study with other state-of-the-art models.

6.9.2.1 Off-line Training

Based on the 0.01 second step size of TDS, the time window width ΔT is set to 0.05 second, so each snapshot of voltage trajectory consists of five points. Considering all the buses in the system, the number of training features is 195 (39 buses × 5 trajectory points). The latest decision time T_F is set to 1 second. Thus $F = 20$ probabilistic predictors are needed.

Two RVFL ensemble models are sequentially trained as the TVSI predictor and the error predictor in each probabilistic prediction model. The number of RVFLs in each ensemble N_R is set to 200, and *sigmoid* is used as the activation function of RVFLs. For the probabilistic prediction performed at each time point, the optimal hidden node range $[h_{min}, h_{max}]$ of the RVFLs is separately tuned, and the tuning result is shown in Fig. 6.26, where the different colors refer to the tuning results for different time points. The hidden node range with the lowest testing MAPE is empirically selected as the $[h_{min}, h_{max}]$ of each probabilistic prediction model. As shown by the arrow in Fig. 6.26, the RVFLs trained for a later time point result in a lower MAPE. Such a phenomenon coincides with the initial thought, i.e. higher prediction accuracy is at the cost of slower decision speed.

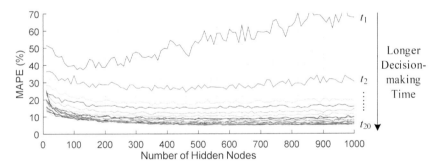

Fig. 6.26: Hidden nodes tuning results

6.9.2.2 Reliability of Confidence Interval

To statistically test the reliability of the predicted CI, the probabilistic prediction models are tested on each testing FIDVR instance at each time point, then 23,280 (1,164 testing cases × 20 time points) CIs are consequently predicted. According to (C. Wan, *et al.*, 2014), the average coverage probability (ACP) index is used to evaluate the reliability of the predicted CIs. Its mathematical formulation is as follows:

$$\mathrm{ACP} = \frac{1}{N_{\mathrm{test}} \times F} \sum_{i=1}^{N_{\mathrm{test}}} \sum_{j=1}^{F} c_{i,j} \qquad (6.39)$$

where

$$c_{i,j} = \begin{cases} 1, & t_i \in CI_j^{\alpha}(t_i) \\ 0, & t_i \notin CI_j^{\alpha}(t_i) \end{cases} \tag{6.40}$$

where N_{test} is the number of testing FIDVR instances, F is the number of time points, t_i is the true TVSI value of the ith testing FIDVR instance, and $CI_j^{\alpha}(t_i)$ is the predicted CI for the ith testing FIDVR instance at the jth time point.

ACP is the average probability of the true TVSI value to fall in the predicted CI. For a reliable CI, its computed ACP should be comparable to its predefined $100(1 - \alpha)$ per cent confidence level. In the test, the ACP of the CIs is 99.07 per cent, which is very close to the predefined 99 per cent CI confidence level. This verifies the high reliability of CI, indicating the high reliability of the probabilistic prediction model.

6.9.2.3 Computational Efficiency

The computational efficiency of the 20 probabilistic prediction models is listed in Table 6.25 where all the tests are performed on a PC with 3.4 GHz CPU and 16GB RAM. Benefiting from the fast learning speed of RVFL, the total off-line training time of 20 probabilistic prediction models is only 3,945 seconds, although each of them includes two ensembles of 200 RVFLs. For on-line testing, the computational efficiency requirement for each probabilistic prediction model is that the computation time consumed by each probabilistic prediction model must be shorter than the time window width ΔT to ensure that there is no time overlap between two successive predictions. In Table 6.25, the average and the longest computation time of the probabilistic prediction on a trajectory snapshot are listed. It can be seen that the longest testing time of 9.88 milliseconds is much shorter than ΔT (50 milliseconds), meaning that the computational efficiency of the probabilistic prediction models is fully compatible with the self-adaptive application.

Table 6.25: Computational efficiency of probabilistic prediction models

Off-line training time	Average testing time on a trajectory snapshot	Longest testing time on a trajectory snapshot
3945 s	7.22 ms	9.88 s

6.9.2.4 Testing Results

Using the developed IS, the FIDVR severity of the 1,164 testing instances is successfully assessed by the 20 probabilistic prediction models. In the test, the applied TVSI threshold $\overline{\text{TVSI}}$ is 1, and the confidence level of the predicted CI is set to 99 per cent.

The testing results are shown in Table 6.26, where the number of successfully assessed instances and the SP accuracy are listed for each time point, and the overall accuracy is also computed. Besides, since the time point to deliver the SP decision varies among different FIDVR instances, average decision time (ADT) is calculated to evaluate the SP speed. The ADT is calculated as the average SP time consumed on each testing FIDVR instance, and smaller ADT indicates higher FIDVR prediction speed.

In Table 6.26, the accuracy at the time points before T_F is 100 per cent, demonstrating the exceptional reliability of the early SP mechanism. It is also important to note that the ADT in the test is only 0.14 second, by a marked contrast to the 1.5 second, 0.35 second, and 0.5 second decision time reported in the literature (S.M. Halpin, *et al.*, 2008; H. Bani and V. Ajjarapu, 2010; Y. Dong, *et al.*, 2017).

Table 6.26: Testing results of probabilistic self-adaptive FIDVR prediction

Time points	No. of assessed instances	Accuracy	Time points	No. of assessed instances	Accuracy
1	793	100%	11	13	100%
2	88	100%	12	5	100%
3	59	100%	13	8	100%
4	39	100%	14	6	100%
5	33	100%	15	3	100%
6	19	100%	16	2	100%
7	26	100%	17	1	100%
8	11	100%	18	2	100%
9	9	100%	19	0	N/A
10	14	100%	20	31	87.10%
Overall accuracy		99.66%	ADT		0.14 s

Moreover, to demonstrate the self-adaptive changing of the predicted CIs, 50 instances are randomly selected from the 1,164 testing instances, and their actual TVSI values and the CIs predicted at T_1, T_{10} and T_{20} are plotted in Fig. 6.27. It can be seen that the CIs at T_1 and T_{20} are the widest and the narrowest, indicating the reduction in prediction error at a later time point.

6.9.2.5 Sensitivity Analysis

In the developed IS, the confidence level and the TVSI threshold are empirically selected. Thus their effects on the FIDVR prediction performance should be further analyzed. With different confidence levels and TVSI threshold values, the SP performance, including the accuracy

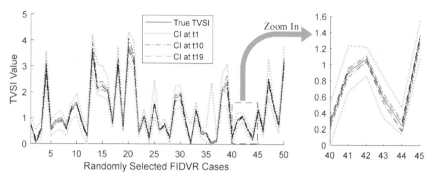

Fig. 6.27: The true TVSI values and the predicted CIs for 50 testing instances

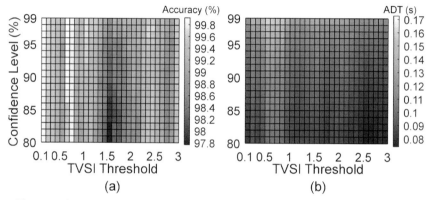

Fig. 6.28: Sensitivity analysis of confidence level and TVSI threshold on (a) SP accuracy, and (b) ADT

and ADT, is demonstrated in Fig. 6.28 where the confidence level varies between 80 per cent and 99 per cent and the TVSI threshold ranges from 0.1 and 3. The worst case SP accuracy and ADT are respectively 97.8 per cent and 0.17 second, which are still recognized as excellent performances for FIDVR prediction. This sensitivity analysis result indicates the robustness of the developed IS to the variation of these two quantities. It is shown that the effect of TVSI threshold on SP accuracy is more significant than the confidence level, and SP efficiency (indicated by ADT) is highly related to the adopted confidence level.

6.9.2.6 Comparative Study

To more clearly demonstrate the advantage of the developed IS, three other intelligent models, including DT, SVM, and Bayesian linear regression (BLR), are selected for comparison. The purpose of comparison with DT and SVM is to demonstrate the advantage of the probabilistic self-adaptive SP structure, and comparison with BLR is to demonstrate the performance of the developed probabilistic prediction model. The tests are performed on the same dataset to make a fair comparison, and the testing results, including the accuracy and the decision time, are shown in Table 6.27.

Table 6.27: Comparative study results for FIDVR prediction

Methods	FIDVR prediction type	Accuracy	Decision time
DT	Fixed-time	99.05%	0.75 s
SVM	Fixed-time	99.66%	0.80 s
The developed IS	Probabilistic self-adaptive	99.66%	0.14 s
BLR	Probabilistic self-adaptive	98.37%	0.33 s

In Table 6.27, the testing results are obtained as follows: for a fixed-time SP method, the highest decision accuracy over different time points is selected as its SP accuracy, and the earliest time point achieving this accuracy is the required decision time. In doing so, the best performance of the fixed-time SP methods is pursued. For a probabilistic self-adaptive IS, the overall accuracy and the ADT over all the testing instances are respectively the SP accuracy and the required time.

There are three findings from Table 6.27:

- The black-box models (i.e. the developed IS and SVM) have higher accuracy than the transparent models (i.e. DT and BLR), confirming the trade-off relationship between the accuracy and the transparency of data-mining tools in (I. Kamwa, *et al.*, 2011). The probabilistic self-adaptive SP achieves shorter prediction time than the fixed-time SP, thereby verifying that the probabilistic self-adaptive IS can significantly improve the FIDVR prediction speed.
- The developed probabilistic prediction model outperforms BLR in terms of FIDVR prediction accuracy and speed.

6.10 Conclusion

This chapter focuses on IS applications on power system post-disturbance real-time SP. In this chapter, four self-adaptive ISs for real-time SP are developed so as to improve the SP speed without impairing the SP accuracy.

The first self-adaptive IS is developed for PMU-based real-time TSP, aiming to balance the response speed and accuracy. The IS is based on the ELM algorithm and ensemble learning techniques with strategically designed learning and decision-making rules. Compared with existing models, it is self-adaptive in order to make the right TSP decision at an appropriate earlier time, hence allowing more time to take the remedial control actions. Numerical test results on the New England 39-bus system and IEEE 50-machine system show that the IS makes the TSP decisions much faster while maintaining a high accuracy. In the meantime, the

IS has a very high learning speed which allows it to be on-line updated for performance enhancement. The developed IS can be used for more efficiently predicting the transient stability status at emergency stage and triggering emergency controls as early as possible to protect the system against catastrophic blackout.

The second IS is developed under a hierarchical and self-adaptive architecture for real-time STVS prediction. Under a hierarchical process, the IS not only detects the voltage instability scenario, but also predicts the RVSI value to quantitatively evaluate the FIDVR severity. Moreover, the self-adaptive SP process minimizes the overall SP earliness without impairing the SP accuracy. The STVS prediction test on New England 39-bus system demonstrates its excellent accuracy and exceptional early SP ability for both voltage instability detection and FIDVR severity prediction. This would significantly enhance the situational awareness of the power system and reduce the risk of blackout triggered by short-term voltage instability.

The self-adaptive IS is then upgraded into a hybrid self-adaptive IS. Unlike previous IS that uses a single learning algorithm, this hybrid IS combines multiple RLAs, including ELM and RVFL, into a hybrid randomized ensemble model. This provides more diversified machine learning outcome. Under an MOP framework, STVS is assessed in an optimized self-adaptive way to balance STVS accuracy and speed. Compared to ISs with single learning algorithm, this hybrid IS could further improve the STVS prediction performance. The simulation results on New England 39-bus system and Nordic system verify its superiority over IS based on single learning algorithm in terms of SP accuracy and speed. In particular, its real-time SP speed is 27.5 per cent~37.3 per cent faster than the single-algorithm-based IS. Given such enhanced SP speed, the hybrid self-adaptive IS contributes to earlier and more timely stability control (e.g. load shedding) for less load-shedding amount and stronger effectiveness.

Furthermore, a probabilistic self-adaptive IS is developed for more efficient, more reliable, and more robust self-adaptive STVS prediction. Compared to the IS previously developed in this chapter, this probabilistic self-adaptive IS do not involve any user-defined parameters, which avoid the high parameter-tuning burden and improve the reliability and robustness of real-time STVS prediction. This non-parametric IS can be presented in two forms – respectively designed for voltage instability detection and FIDVR prediction. The numerical tests on New England 39-bus system demonstrate that, by using the probabilistic self-adaptive IS on real-time STVS prediction, 100 per cent accuracy can be achieved at earlier time windows, meaning that the overall voltage instability detection speed and the FIDVR prediction speed reliably improve without increased miss alarm.

Intelligent System for Emergency Stability Control

7.1 Introduction

As the previous chapters have gone through the IS in power system on-line SA, preventive control, real-time SP, this chapter focuses on the last action, i.e. emergency control to complete the SA&C loop in Fig. 2.3. In this emergency situation, there are two streams of control measures to avoid system instability. The first stream is to improve the transmission capability through active/reactive power control, such as high-speed excitation, power system stabilizer, VAR compensation, etc. (P. Kundur, *et al.*, 1994; A.S. Meliopoulos, *et al.*, 2006; J. Liu, *et al.*, 2017). These measures require permanent reinforcement of the grid and have limited control capability against severe disturbances. By contrast, the other stream is to activate the emergency control actions, such as generation adjustment and/or load shedding, which are effective forces to suppress the power oscillation and recover stable system operation. Emergency control is a remedial action in SA&C, and is often seen as the last resort to maintain power system stability.

As one of the three categories of dynamic security, frequency stability refers to the ability of a power system to maintain steady frequency following a severe system upset, resulting in a significant imbalance between generation and load (P. Kundur, *et al.*, 2004). Frequency instability may appear as sustained frequency swings and/or leading to tripping of generating units and/or loads. Generally, frequency instability is caused by a significant sudden imbalance between power generation and load of the system, e.g. an unwanted outage of a generator. Control actions to counteract frequency instability include load shedding and generator tripping, where the former reacts for the generation loss while the latter responds for the load loss. Both aim to retrieve the power balance of the disturbed system.

As an effective emergency control action, load shedding for frequency instability prevention can comprise response-driven and event-driven strategies. For response-driven load shedding, the load shedding is triggered by system frequency response following disturbance(s), known as underfrequency load shedding (UFLS) (Y.-Y. Hong and S.-F.Wei, 2010; L. Sigrist, *et al.*, 2009; X. Lin, *et al.*, 2008; C.-T. Hsu, *et al.*, 2005). When the system post-fault frequency drops below pre-determined thresholds, the UFLS devices shed a certain percentage of loads according to the settings. By contrast, event-driven load shedding (ELS) is automatically activated by a recognized event (contingency), and trips load immediately (based on pre-simulations) without waiting for observing frequency declines (W. Da, *et al.*, 2009). For enhanced power system frequency stability, both load shedding schemes are necessary in practice, and they should be appropriately coordinated in practice. Generally, the ELS is designed for very severe events with high probability; UFLS, on the other hand, is applied for less likely contingencies. In Chinese power systems, the ELS strategy has been implemented at the second and UFLS at the last line of the space-time cooperative framework for defending wide-spread blackouts (Y. Xue, 2007). This chapter focuses on ELS.

Historically, the calculation of ELS is based on deterministic approaches, which heavily suffer from intensive computation burden and mismatching problems. To overcome and/or supplement the conventional methods, this chapter proposes an IS methodology for enhanced ELS against severe contingency events. The methodology consists of off-line training of a fast and accurate prediction model, based on ELM and on-line applying it for predicting the required ELS given a contingency.

7.2 Load Shedding and Its Strategies

Since frequency instability is essentially caused by the significant active power imbalance in the power system, it is feasible to relieve the frequency decline and restore it to an acceptable value by fast retrieving the generation-load balance. The most efficient actions are load shedding (for generation lost situation) and generator tripping (for load demand loss situation). In addition, another effective measure is the fast activation of spinning reserve. However, it is often helpless for very severe disturbances, e.g. a large amount of generation is lost, due to its limited capacity and slow-acting speed. In this chapter, the focus is only placed on load shedding. According to different intervention timing and logic, load shedding includes two strategies as follows:

- **Event-driven load shedding**

The ELS is driven by a contingency (event). Once the contingency is detected through the recognition of the status of system elements, it immediately sheds the loads without waiting for system frequency to start decline. The process is conceptually described in Fig. 7.1.

• **Under-frequency load shedding**

The UFLS is driven by system post-disturbance frequency trajectory, when the frequency drops to specified values (such as 49.10 Hz or below). It then trips a certain percentage of loads at each stage. The process is conceptually described in Fig. 7.2.

Generally, the two load shedding strategies are essentially different in response logic, control effectiveness, cost, and complexity of determination, etc.

Fig. 7.1: Process of the ELS

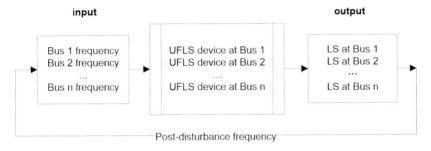

Fig. 7.2: Process of the UFLS

7.3 Response Logic and Intervention Timing

The ELS strategy is implemented based on a decision table featured by input and output, where the input includes the quantities characterizing the current system OP and the contingency characterized by the status of system elements. The output is the load shedding location and the amount. Once the input is fed, the output can be immediately indexed and the corresponding load shedding actions are triggered. So there is almost no time delay after the contingency occurs (although a very short delay due to the communications and relay actions is required but it can be neglected as it is normally within 10 ms). The ELS is feedforward in control logic; it only reacts one time for the detected event, and does not require further input despite the system frequency fluctuations.

By contrast, the UFLS counteracts frequency decay after the frequency drops already below predetermined thresholds. Its input is local frequency measurement and the output is the local load-shedding actions determined by its settings. The settings include the total number of load-shedding stages, the percentage of load to be shed at each stage, and the time delay for each stage. The UFLS is feedback in control logic since it requires continuous input (real-time frequency measurement) to lead its actions. The ELS is centralized since the control signal is sent from the control center (where the decision table is located), and UFLS is decentralized since the control signal is local information.

7.4 Control Effectiveness

In terms of control efficiency, ELS is considered much higher than that of UFLS. The reason is simple: firstly, ELS acts much earlier; thus it can resist the frequency decay in advance so as to recover system frequency back; secondly, ELS is implemented at system level to retrieve the balance between generation and load. By contrast, UFLS is based on local information and it only takes effect for local frequency recovery without using the whole information of the system frequency response.

In terms of control accuracy, UFLS could outperform ELS since it takes action according to actual frequency trajectories, whereas ELS is based on the decision table which may suffer from non-negligible uncertainties in system modeling and system's operating condition as well as load characteristics during its off-line preparation. Besides, it is very challenging computationally, if not impossible, for ELS to include all the possible contingencies, especially due to the power market and renewable energy which introduce huge uncertainties. Consequently, there is the possibility for the ELS decision table to mismatch the fault in its practical application, which would lead to ineffectiveness to prevent the frequency instability.

In terms of control cost, it is usually the case that ELS requires less load-shedding amount than that of UFLS to recover frequency to the same level or FSM, and the under-frequency duration time (which is another important factor in evaluating the cost of the frequency emergency control) under ELS strategy can be smaller than that of UFLS. Besides, ELS may only involve a small number of load-shedding locations while UFLS could affect a wider area due to its local control logic.

Two simulation examples are given in Figs. 7.3 and 7.4 to explain this, where the two load-shedding strategies are respectively illustrated. For UFLS strategy, simulation shows that the earliest load shedding is

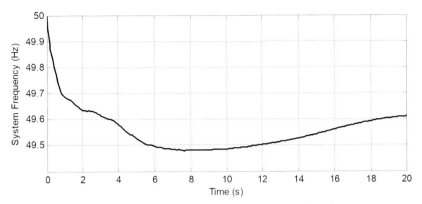

Fig. 7.3: Post-fault frequency trajectory with ELS

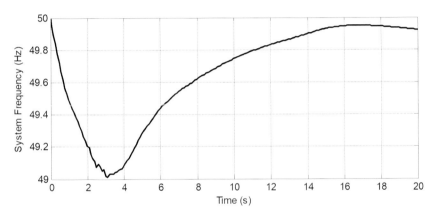

Fig. 7.4: Post-fault frequency trajectory with UFLS

triggered at 3 s, and since then, all the UFLS devices have to counteract the frequency decline. The final total UFLS amount is 1699.5 MW, and the frequency after 20 s is 49.92 Hz. For ELS strategy, it can be noticeably seen that after its fast intervention (at 0.2 s), the system frequency decline becomes much slower and smoother. The final total ELS amount is 905.2 MW, and the frequency after 20 s is 49.63 Hz.

Comparing the two strategies, in terms of control efficiency, ELS is considered much higher than that of UFLS for the reason that ELS has a quicker response to frequency decline. Thus, ELS has a shorter under-frequency duration time (say, below 49.5 Hz) than that of UFLS. In terms of control cost, ELS is much cheaper than UFLS. In terms of control accuracy, UFLS could outperform ELS since it takes action according to actual frequency trajectories. So the ultimate frequency is closer to nominal value (say, 50 Hz) than that of ELS.

7.5 Computation Complexity

The calculations of the two load-shedding strategies are significantly different, where ELS is much simpler than UFLS.

For ELS, as already mentioned, the decision table is critical since it is responsible for on-line determining the required load-shedding action. The preparation of the decision table consists of simulations with respect to imagined contingencies on the given system operating conditions. Traditionally, the preparation of the decision table is performed at off-line stage. The OPs are forecasted based on load forecasting and generation dispatching information. Once the decision table is finished, it is used on-line without updating, which is termed as 'off-line simulation on-line matching'. However, such off-line approach may heavily suffer from inaccuracies due to the uncertainties and approximations in OP forecasting and system modeling. Even more importantly, the off-line approach may be unable to cover the practical system conditions, which may lead to ineffectiveness or over-shed problems. With the advancement of computing hardware and powerful analysis tools, such as FASTEST (Y. Hue, 1998) being available to perform on-line stability analysis, it becomes feasible that the ELS decision table be on-line generated using practical operating information and be updated periodically to catch the varying system conditions. This is termed as 'on-line calculation real-time matching' (Y.J. Fang and Y. Xue, 2000). Such computation architecture can significantly enhance the accuracy and robustness of the decision table.

On the other hand, the UFLS actions depend on its settings, which are manipulated on the UFLS devices in field, and once set, should be fixed during its application. Consequently, unlike ELS decision table which couples one load-shedding action with one OP/contingency pair, the settings of UFLS should cover all the OPs and the contingencies considered. Besides, the settings of UFLS are performed off-line and cannot be updated on-line.

In calculating the two load-shedding strategies (decision table for ELS and settings of UFLS), the problem is regarded as an optimization model which can be formulated as minimizing an objective function subject to given constraints:

$$\text{Minimize} f(x, u, o, c) \tag{7.1}$$

$$\text{s.t.} \quad g(x, u, o, c) = 0 \tag{7.2}$$

$$h(x, u, o, c) \leq 0 \tag{7.3}$$

where u and x respectively represents the vector of controllable and dependent variables; c is the set of contingencies to be considered, and o is the system OPs under calculation; $f(x, u, o, c)$ is the objective function, and usually represents the cost or the load-shedding amount, or even their combination (in this case, the optimization is a MOP problem); $g(x, u, o, c)$ is the set of equality constraints (such as power flow equations), and $h(x, u, o, c)$ represents the set of inequality constraints, where the frequency stability requirement is included.

However, the calculation for the two load-shedding strategies under the above model is significantly different in complexity and difficulty. Some critical differences are as follows:

- **Control variables**, i.e. u: ELS may only need to consider the load-shedding location and amount, whereas UFLS needs to comprise load-shedding stages, time delay for each stage, percentage of load allowed to be shed in each stage, etc., which are much more complex than that of ELS.

- **Contingency list and OP**: ELS decision table associates one load-shedding action with one OP/contingency pair. So it decouples o and c in optimizing u; in other words, it calculates the load-shedding scheme for one OP and one contingency. By contrast, UFLS settings need to cover as many scenarios as possible. So it collectively considers the OP and contingency list in its calculation. However, since the system OP can be widely varied, it is usually very difficult to reach the tradeoff between the OP and the contingencies. In literature, the representative OPs and contingencies are used to simplify the problem (L. Sigrist, *et al.*, 2009). However, it should be pointed out that the UFLS may fail to prevent the frequency instability, or over-shed the load in other situations due to its imperfect considerations in selecting OP and contingencies.

- **Solution algorithms**: Since ELS has only two optimization parameters, with the use of FSM, sensitivity-based approach can be used to solve the model iteratively (W. Da, *et al.*, 2009); but for UFLS, the optimization parameters and system frequency stability are complicatedly related, making the problem highly non-linear. Thus it is much more difficult to solve the model by classic programming techniques, but computational intelligence algorithms could be an alternative to get higher-quality solutions (Y.-Y. Hing and S.-F. Wei, 2010).

- **Computational efforts**: It is also worth mentioning that computational efforts are required by the two load-shedding strategies since ELS needs to be performed in an on-line environment, while UFLS is conducted in off-line stage. Apparently, the ELS strategy needs more computing and real-time communication sources.

7.6 Hardware and Infrastructure

The two load-shedding strategies have different hardware requirements for their deployments in practice.

The ELS is a centralized control scheme; so a reliable communication link and associated sensors/measurement units which are responsible for collecting and transmitting the signals between the control center and the load-shedding devices are needed. To enhance the communication reliability, the redundancy in communication link is also critically required.

On the other hand, the UFLS is activated in a decentralized way, so that the load-shedding devices only need local sensors/measurement units.

7.7 Coordination between ELS and UFLS

It is realistic to admit that neither of the two load-shedding strategies can perfectly handle all the risk of frequency instability because of the unlimited number of possible scenarios and increasing uncertainties in today's power systems. How the two load-shedding strategies are coordinated in power system operation is focused in this section. The benefits, difficulties and possible solutions involved in this coordination are analyzed. A vision on the future advancement of load-shedding strategies is also provided at the end of the section.

7.7.1 Benefits

The ELS, on one hand, is strong enough to rapidly retrieve system frequency by fast load-shedding actions, but may suffer from limited contingency consideration and corresponding mismatching problems, as well as unsatisfactory control accuracy due to the uncertainties. The UFLS, on the other hand, is accurate and robust, but is relatively weak in control efficiency due to its response-based intervention mechanism. Consequently, the two load-shedding strategies can be complementary to each other, and the appropriate coordination between them can, not only enhance the control effectiveness, but also reduce the overall control cost. Generally, the ELS should be designed for those events with large certainty, whereas UFLS should be resorted to for relatively uncertain events.

Three simulation examples are shown in Figs. 7.5 to 7.8 to illustrate the need and the benefit of such coordination. Note that a new OP, which is more stressed than the one above, is used for the illustration here.

Under the same contingency with Fig. 7.3, it can be noted from Fig. 7.5 that the resulting post-disturbance frequency has dropped to 47.381

Hz, and the FSM is –71.11. In Fig. 7.6, the ELS (also 905.2 MW) is taken to resist the frequency decline. However, the control is inadequate since the frequency is only 48.69 after 20 s and the FSM is –9.23. On the other hand, in Fig. 7.7, the UFLS can prevent frequency instability under this new OP but with a much more expensive control cost (2087.1 MW). The system frequency with UFLS after 20 s is about 49.84 Hz.

The cooperation of the two load-shedding strategies is illustrated in Fig. 7.8, where it can be seen that the frequency after 20 s is 49.31 Hz. The final total ELS is still 905.2 MW and final total UFLS is 493.9 MW. Consequently, the total cost is much less than the single UFLS, and the system frequency is stable as compared to the single ELS.

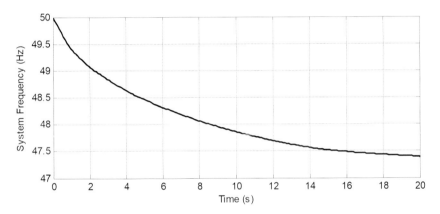

Fig. 7.5: Post-disturbance frequency trajectory without load shedding (a stressed OP)

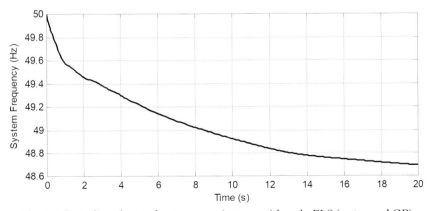

Fig. 7.6: Post-disturbance frequency trajectory with only ELS (a stressed OP)

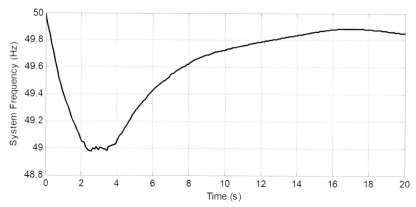

Fig. 7.7: Post-disturbance frequency trajectory with only UFLS (a stressed OP)

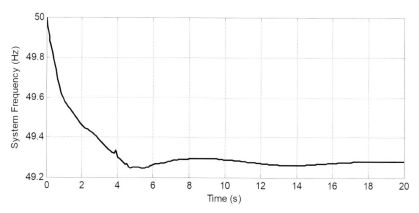

Fig. 7.8: Post-disturbance frequency trajectory with both ELS and UFLS
(a stressed OP)

7.7.2 Difficulties and Possible Solution

Although promising, the exact coordination of ELS and UFLS is however very intricate since it is a highly non-linear, non-convex, and high dimensional mixed integer programming problem.

Generally, the problem can be modeled as follows, but the optimization parameters for the two load-shedding strategies are coupled:

$$\text{Minimize} \quad F[(x_1, u_1, o_1, c_1), (x_2, u_2, o_2, c_2)] \tag{7.4}$$

$$\text{s.t.} \quad G[(x_1, u_1, o_1, c_1), (x_2, u_2, o_2, c_2)] = 0 \tag{7.5}$$

$$H[(x_1, u_1, o_1, c_1), (x_2, u_2, o_2, c_2)] \le 0 \tag{7.6}$$

where subscript 1 denotes variables for ELS and 2 denotes variables for UFLS, respectively.

Since the two load-shedding strategies take effect at different timings, conventional objective functions which only consider load-shedding amount or frequency deviation may become unsuitable. It is therefore important to associate the two different load-shedding strategies within one objective function.

Besides, since the two load-shedding strategies are wholly coupled during their practical actions, it is also very intricate to directly solve the model. One possible way is to decouple the two load-shedding strategies during the solution (X. Yusheng, *et al.*, 2009), and another way is to rely on modern heuristic optimization techniques, such as EAs for solution.

7.7.3 Envisions

In addition to the coordination of two load-shedding strategies, with the continued advancement of computing power and communication ability, such as WAMS, in the smart grid environment, the two load-shedding strategies can be upgraded significantly. Some envisions are given below:

Centralized UFLS: Unlike conventional local UFLS, centralized UFLS scheme can utilize the system-wide information, such as frequency at remote buses, to make load-shedding decisions (load-shedding locations, stage, time delay, and the amount). With the entire system information, the centralized UFLS can preserve frequency stability more efficiently, especially after large disturbances.

Advanced computing technique-assisted ELS: Due to the parallel nature of the ELS calculation, advanced computation technique, such as distributed computing and cloud computing, can significantly speed up ELS on-line computations.

IS-assisted ELS: As a universal tool, IS technique is also prospective to assist deterministic approaches for fast calculation of ELS. IS technique has a strong capability of non-linear modeling of any complex mapping relationship. It has been shown promise to facilitate real-time SP. So it can be also utilized to predict the ELS which has an input-output structure. This idea is further extended in this chapter by developing an ELM-based model for real-time prediction of ELS.

Demand response-based load shedding: In smart grid context, demand response will play a key role in system operation and control, especially frequency control (Z. Xu, *et al.*, 2010). Demand response will be available in both steady state as well as emergency conditions to meet specific objectives. A recent reference (Y. Wang, *et al.*, 2011) has reported a method for voltage collapse prevention with demand response. Consequently, it can be envisioned that the availability of demand response could also provide significant benefits in design and application of frequency stability load shedding.

7.8 State-of-the-Art of ELS

In practice, the ELS decision table is prepared based on TDS and associated analysis. Historically, the process is conducted in the off-line stage, where prospective OPs are forecasted and contingencies are imagined at specific lead time. For each OP/contingency scenario, the ELS (if needed) is calculated with respect to given limits and frequency stability requirements. However, with the introduction of open power market and increased penetration of renewable energies, such as wind power, modern power systems become uncertain and highly unpredictable, making the conventional off-line approach insufficient.

Based on the continuous advance of computing hardware and algorithms, it is now feasible to move the off-line calculation to on-line stage, where the ELS decision table can be refreshed by current on-line operating information. This is termed as 'on-line pre-decision' (Y.J. Fang and Y. Xue, 2000). This scheme can significantly enhance the accuracy and robustness of the decision table since it can track the practical power system operating states.

However, due to the deterministic nature, both off-line and on-line approaches can only accommodate a limited number of contingencies. In the face of unlimited possible event scenarios, especially combinational events, the decision table by deterministic approaches could suffer from mismatching problems, and if so, the power system may lose frequency stability, when UFLS, as the last line of defense, fails to counteract. In particular, renewable energy, especially wind power generation, is widely installed in modern power systems over the recent years. Due to the intermittent and stochastic nature of wind resource, wind power generation is highly variable and uncertain, which can result in various power-imbalance scenarios. For instance, there may be complex generation loss patterns. Consequently, there is a pressing need for more efficient tools to fast determining required ELS in case of a large severe contingency. The developed IS based on ELM can meet this demand as it can be real-time applied for ELS prediction and can generalize the prediction in both seen and unseen scenarios.

7.9 IS for ELS Computation

To overcome the inadequacies of the deterministic approach for ELS, this chapter presents an IS that can universally predict the required ELS against severe contingencies. The general idea of the methodology is introduced as follows:

Similar to preparing the decision table, a comprehensive ELS database is firstly generated based on various OP/contingency simulations and corresponding optimal ELS calculations. The optimal ELS means the minimum ELS to maintain frequency stability following a contingency. It is calculated based on a quantitative frequency stability index and an iterative optimization process, which will be introduced in detail in subsequent sections. Unlike deterministic schemes, in the developed IS, a contingency is characterized by the resulting total loss of generation (MW), total loss percentage of generation (%), and total generation-load imbalance (MW), whereas an OP is defined by active output of each generator (MW) and active power load of each bus (MW). As a result, the contingency and OP can be generally detected rather than deterministically identified.

Secondly, the knowledge (i.e. the mapping relationship between OP/contingency scenarios and the required ELSs to maintain frequency stability) is extracted by a machine learning model based on ELM. Once the ELM is well trained, the prediction model can be applied to predict the required ELS in real-time. Due to the strong generalization capacity of ELM, both foreseen and unforeseen scenarios can be predicted. Benefiting from the computationally efficient training speed of ELM, the prediction model can be on-line updated/enriched by on-line deterministic calculations. The prediction model can be an individual tool or statistical complement to the existing on-line tools to enhance the overall reliability of ELS strategy.

It is also worth mentioning that in (C.-T. Hsu, *et al.*, 2005), the authors propose a strategy that applies ANN to determine the load shedding, which is similar to the idea in this chapter in terms of relying on machine learning technique for load shedding. However, the approach reported in (C.-T. Hsu, *et al.*, 2005), is designed for response-driven load shedding, while the method in this chapter is for event-driven load shedding. Even more important, the developed IS adopts a quantitative frequency stability criterion and optimization approach for generating the database. Besides, this chapter employs ELM technique for capturing the input-output relationship. The ELM is advantageous over conventional ANN algorithms as it has much faster learning speed, better generalization capacity, and efficient tuning mechanism, which can thus allow effective on-line updating for enhanced prediction. Furthermore, the developed IS in this chapter can also be a supplement to conventional deterministic tools.

7.9.1 Quantitative Frequency Stability Index

Frequency instability by a large disturbance may appear as sustained frequency swings, decay or rise. To design economic load-shedding strategies, the first step is to evaluate the severity of the phenomenon. In the literature, the lowest swing frequency (Y.-Y. Hong and S.-F. Wei, 2010), frequency decline ratio (C.-T. Hsu, *et al.*, 2005), generation deficiency and frequency decline ratio (GD/FD) (R.F. Chang, *et al.*, 2005), are employed to indicate the severity of frequency decay and the required load-shedding strategies are calculated based on such indices. However, these indices can only measure the severity of the disturbance at only one time point and thus, are insufficient to measure the frequency violation along with its affecting period to consumers (where the frequency violation time experienced by each bus should be considered). Besides, they are unable to reflect different frequency deviation acceptability of different buses (consumers). Further, these indices are not continuous and thus cannot quantitatively assess the frequency stability degree.

For the developed IS, the FSM introduced already in Section 1.7.3 is adopted for quantitative evaluation of frequency stability severity. The advantage of FSM over other indices is obvious, as it can overcome the deficiencies of the above-mentioned indices. Even more valuable, sensitivity methods can be employed to efficiently search the stability limit and/or optimize the required controls for instability conditions.

7.9.2 Computation of Optimal ELS

In practice, the selection of load-shedding buses involves the power supply priority, operating rules, as well as human judgement. When it comes to minimizing the required ELS for restoring the frequency stability to a desired level, the problem can be formulated as:

$$\min \ \sum_{i=1}^{m} \Delta L_i \tag{7.7}$$

$$\text{s.t. power flow equations} \tag{7.8}$$

$$\text{static operating limits} \tag{7.9}$$

$$0 \leq \eta \leq \varepsilon \tag{7.10}$$

where ΔL_i is the load shedding amount at permissible bus i and m is the total number of permissible bus; static operating limits (7.9) consist of generation output limits, bus voltage limits, and transmission line thermal limits; η denotes the FSM and is a function of ΔL_i, $\eta = F(\Delta L_1, \Delta L_2, \cdots \Delta L_m)$, ε is a user-defined threshold which represents the desired control effect (note that a larger η requires a larger load-shedding amount, so there is need to constrain η by an appropriate threshold to avoid over-control).

It is evident that the above model consists of a non-linear programming problem. Because of the implicit nature of function F, the model cannot be solved directly; instead, a sensitivity-based iterative solution process can be employed (Y.-H. Chen and Y.-S. Xue, 2000). The idea is to linearize the above model at small steps, and at each step selects the most effective bus to apply the ELS, thereby increasing the FSM step by step. At the last step, the FSM is improved to the desired level. The whole computation process is as follows:

(1) Given a supposed OP/contingency scenario, perform TDS to calculate the initial FSM η^0, if η^0 satisfies (7.10), stop; otherwise, let $k = 1$, go to step (2);

(2) At iteration k, for bus i, $i = 1,2,\ldots,m$, calculate its FSM sensitivity with respect to a small load-shedding step λ (e.g. 5 MW) by two successive TDS runs, $S_i^k = \dfrac{\Delta \eta^k}{\lambda_i} = \dfrac{\eta^k - \eta^{k-1}}{\lambda_i}$;

(3) Rank the buses according to their sensitivity value, and apply the ELS at the largest bus at this iteration, $\Delta L_i^k = \Delta L_i^{k-1} + \lambda_i$;

(4) Recalculate the FSM η^k with the ELS ΔL_i^k. If the FSM η^k satisfies (7.10) and other constraints of (7.8) and (7.9) are also met, stop; otherwise, let $k = k + 1$ and return to step (2).

7.9.3 ELS Database

The ELS database is characterized by features and objects, where features are system pre-fault OP and the contingency (event) information, and the objects are corresponding optimal ELS (subject to the frequency stability requirement).

Since power system frequency stability is strongly related to the active power balance of the system, in this chapter, we select steady-state system power generation and load variables as features to characterize an OP, and the contingency events are represented by the resulting active power imbalance. The features and objects defined for the ELS database are described in Table 7.1, where the first four items correspond to the OP and the last three items correspond to the contingency event.

Table 7.1: ELS database

	Features	Objects
OP	Active output of each generator (MW)	Optimal ELS (MW)
	Active power load of each bus (MW)	
	Total active power generation (MW)	
	Total active power load (MW)	
Contingency	Total loss of generation (MW)	
	Total loss percentage of generation (%)	
	Total generation-load imbalance (MW)	

7.9.4 Implementation

Once the ELM is properly trained by the ELS database, it can be on-line applied to predict the required ELS in real-time just like a generalized decision table.

In addition to working individually, the prediction model can also be a complement to deterministic tools, which means if an occurred contingency is considered in the deterministic method, the required ELS can be indexed in the decision table; otherwise, the ELS can be predicted by the ELM model. This concept is described in Fig. 7.9. Benefiting from the very fast training speed of ELM, the information in the decision table can also be used to update/enrich the ELM for enhanced prediction performance.

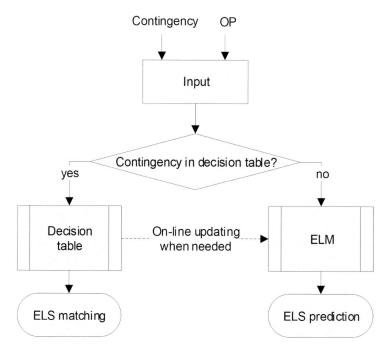

Fig. 7.9: Prediction model as a complement to the deterministic tool

7.10　Simulation Results

The developed IS is tested on the New England 39-bus system. For a comprehensive test, the tenfold cross validation criterion is adopted. To measure the accuracy, the mean absolute error (MAE) and MAPE are used.

7.10.1　ELS Database Generation

A total of 15 OPs are simulated for the test system, which cover a wide range of load/generation patterns (the total load is changed between 50 per cent and 120 per cent of its base level). Four types of N-2 combinational generation loss events are assumed as the contingency events (*see* Table 7.2); for each type of event, 10 different severe disturbances resulting in different generation loss amount between 300 MW and 600 MW are considered. In doing so, 600 instances are produced in total.

Table 7.2: Contingency events

Event No.	1	2	3	4
Generation loss	G2 & G3	G3 & G5	G4 & G5	G3 & G4

FASTEST software (Y. Hue, 1998) is employed to perform TDS where FSM can be directly obtained. Here, the two-element table is set to [(49.00 Hz, 10.00 s)] for each bus of the system. It should be noticed that in practice, the two-element table is set according to specific system operating rules/stability requirements, and thus can vary among different utility companies. TDS results show that the FSM values of the 600 instances are all negative (ranging from –11.9 to –88.2), which means they are all unstable.

In this test, the frequency stability requirement after ELS is assumed as 1.00 ± 0.05 in terms of FSM (in practice, it can be larger for higher frequency stability requirement; however, this may need more load-shedding amount). Bus 25, bus 20 and bus 4 are selected as the permissible load-shedding buses. The corresponding optimal ELS for each instance is calculated based on the method introduced above. It is found that for the 600 unstable instances, the corresponding optimal ELSs range from 51.0~305.0 MW.

7.10.2 ELM Tuning

Training an ELM consists mainly of selecting the activation function and the optimal hidden nodes, where four types of activation functions are available, including *sigmoidal*, *sine*, *hardlim*, and *radial basis* functions. As already mentioned, the optimal parameters can be determined via a tuning process, where the training data is divided into two non-overlapped data sets – one for training and the other for validation. The parameters are selected as the ones resulting in the lowest validation error.

Since there are 600 instances generated in total, in order for an unbiased test, during the tuning, only 400 instances are randomly picked up for use. The 400 instances are randomly divided into a training set (300 instances) and a validation set (100 instances). The ELM tuning profile is given in Fig. 7.10, where it can be seen that *sigmoidal*, *sine*, and *radial basis* functions provide a better performance, and when the hidden node number exceeds 80, the validation accuracy by them are approximate; in particular, the lowest validation error is 2.64 per cent, which is achieved when the hidden nodes are 107 when using *sigmoidal* function.

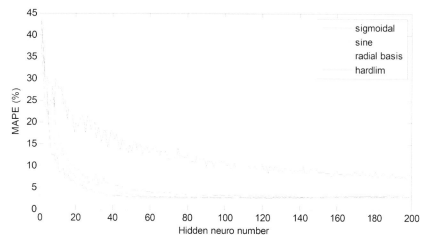

Fig. 7.10: ELM tuning profiles

7.10.3 Test Result

Given the tuning result, the ELM-based prediction model is then tested by all the 600 instances under tenfold cross test criterion. The results are given in Table 7.3. Note that the training and testing time is the total computing time for all the tenfolds.

Table 7.3: Tenfold test results (four events)

Training time (s)	Testing time (s)	MAPE (%)	MAE (MW)
0.312	0.001	3.050	4.837

From Table 7.3, it should firstly be observed that the ELM works very fast as the testing and training time are negligibly small, which means that the model can be real-time applied and on-line updated; secondly, the prediction is very accurate as the tenfold MAPE is only 3.010 per cent and MAE is 4.384 MW, which means the prediction model can effectively predict the required ELS.

To examine the impact of the prediction error on the practical ELS effectiveness, an unseen instance is randomly picked out from the testing set. Its ELS amount is respectively deducted and added the MAE value 4.837 MW and the resulting post-fault system frequency trajectories are obtained via simulation. Figure 7.11 shows the trajectories. The frequency trajectory without load shedding actions is also given in this figure for observing the strong control effectiveness of ELS.

It can be seen from Fig. 7.11 that, with either positive or negative average prediction error included, the resulting frequency trajectories are

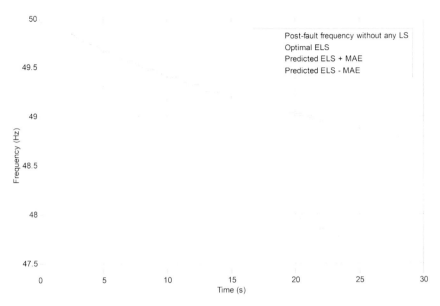

Fig. 7.11: Impact of the prediction error on control effect

approximate to that of the optimal ELS within very limited deviations. In other words, there is no significant impact by the prediction error on the control effect in practical implementation.

7.10.4 Test on Unlearned Events

In the above test, the capability of the prediction model in predicting learned events has been verified with very fast computing speed and satisfactory accuracy. In this test, its generalization capacity on unlearned events is examined. Given the four types of events, the ELM is trained with three types of events and tested on the other one type of event. The results are listed in Table 7.4.

Table 7.4: Test results on unlearned events

Training event	2, 3, 4	1, 3, 4	1, 2, 4	1, 2, 3	Average
Testing event	1	2	3	4	
MAPE (%)	5.69	2.92	3.02	6.12	4.44
MAE (MW)	8.50	3.97	4.24	10.65	6.84

According to the results, the prediction accuracy on unlearned events is less satisfactory: in average, the MAPE and MAE on unlearned events are 4.44 per cent and 6.84 MW. This implies that the model is relatively weaker in generalizing on contingencies than on OPs. However, it is important to remember that conventional decision table could be unable to deal with mismatch problems.

In practice, to enhance the performance of the prediction model, one way is to incorporate as many contingency events as possible during the off-line training stage. The other way is to perform on-line updating/ enriching which will be presented next. Note that although this would result in a larger database for learning, due to the very fast learning speed of the ELM, such off-line and on-line training tasks can be efficiently achieved.

7.10.5 Performance Enhancement by On-line Updating/ Enriching

Benefitting from the very fast learning speed of ELM, the prediction model can be on-line updated/enriched by the on-line refreshed decision table. As already mentioned, this can be an effective way to enhance the performance of the prediction model. In particular, the model can be enriched by new contingency events that are not learned in the off-line stage. To test the feasibility and the accuracy, a new ELS database is generated to enrich the ELM. The database is on a single event, which trips generator 8, and consists of 360 instances. It is found that under this single fault, the required optimal ELS ranges from 102 to 258 MW.

The new 360 instances are attached to the initial database, resulting in an enriched database composed of 960 instances where single and combinational faults are both included. The prediction model is then tested under tenfold test criterion and the whole testing results are given in Table 7.5.

Table 7.5: Tenfold test results (enriched database with five events)

Training time (s)	Testing time (s)	MAPE (%)	MAE (MW)
0.780	0.0245	2.550	4.189

According to Table 7.5, it is important to find that although the database increases in size, the required training and testing time by the ELM remains trivial, which confirms the feasibility of the on-line updating. Even more importantly, it should be observed that after the enriching, the prediction is accurate and is even improved slightly, indicating that the prediction model, when trained with different types of contingency, can generalize its prediction on these contingencies.

It should be noted that in practice, the number of new instances that can be generated on-line for updating depends on system scale and computing hardware, on an ordinary computer with 3.1 GHz CPU as adopted in this chapter. The ELS calculation for one instance costs 5 s using FASTEST software.

7.11 Conclusion

Load shedding against frequency instability comprises response-driven load shedding (known as UFLS) and ELS. UFLS, on one hand, can react to all faults as the last line of power system defense, but with high control cost and low efficiency. ELS, on the other hand, is strong enough to rapidly retrieve system frequency due to its fast intervention mechanism, but it may suffer from mismatch problems.

This chapter first gives a comprehensive introduction and comparison of the two load-shedding strategies and then develops an IS for real-time ELS prediction, using extreme learning machine (ELM) technique. The developed model can be an individual tool to fast predict the required ELS for counteracting various severe contingency events. Besides, it can also be a complement to conventional deterministic approaches to overcome the latter's shortcomings.

The IS is demonstrated on the New England 39-bus system. Based on the simulation results, the following conclusions are made:.

(1) The ELM-based IS works very fast as it requires only trivial testing and training time, which enables its real-time application and on-line updating/enriching.
(2) The prediction accuracy is very high on the learned contingency events and investigation shows that the prediction error has only very minor impact on the control effect.
(3) The prediction accuracy is relatively less satisfactory on unlearned events, which means that we need to incorporate as many contingency events as possible in the off-line training stage and/or perform on-line updating/enriching to incorporate more contingencies.

Moreover, the on-line updating/enriching is an effective means to improve and maintain the practical performance of the IS.

8

Addressing Missing-data Issues

8.1 Introduction

In the literature, IS has been identified as a promising tool for on-line and real-time power system SA&C. Most of the existing ISs (including the ISs established in previous chapters) assume that the wide-area measurements collected in real-time and provided to ISs are complete and accurate. However, in practice, the measurement data may not always be completely available due to unintentional events, such as PMU failure, PDC failure, communication congestion, or even cyber-attack, etc. According to the WAMS reliability analysis in (Y. Wang, *et al.*, 2010), the WAMS reliability of a 14-bus system is estimated as 97.35 per cent, meaning there is a 2.65 per cent chance for data missing. Considering a larger realistic power grid, the probability of data missing would be even higher. When the ISs are exposed to incomplete PMU data, their SA performance can be significantly impaired due to the data-driven nature of their computation mechanisms. Therefore, to be practically sound, the missing-data issue must be considered in designing the IS for power system SA&C.

In data analytic area, various imputation methods have been proposed in the literature to handle the missing-data issue. Some examples are mean imputation, maximum likelihood imputation, and multiple imputation (A.N. Baraldi and C.K. Enders, 2010). Although imputation method can substitute the missing values in the dataset, it induces large computation error compared to the case with a complete dataset, especially when the missing feature is highly effective to solve the problem (M. Saar-Tsechansky and F. Provost, 2007).

Since the large number of components in power grids are usually interconnected, the network's operating features measured in a wide-area manner are usually interdependent. Thus another approach of handling incomplete measurement data at the on-line stage is to intelligently estimate the missing operating features using other available features. In doing so, complete input data will be available to the ISs under missing

PMU measurement conditions. However, this method needs to train a large number of intelligent models to work for all possible feature missing scenarios, which is computationally burdensome or even infeasible. For example, in (Q. Li, *et al.*, 2019) , the feature estimation method can predict the missing data directly, but it needs to train a large number of classifiers for different PMU missing conditions which would suffer from the curse of dimensionality.

More recently, two DT-based methods have been proposed to handle the missing-data issue in on-line SA. One of them is the surrogate DT model proposed in (T.Y. Guo and J.V. Milanovic, 2013). It replaces the DT splitting rule on each missing input by a surrogate splitting rule that is greatly associated with the original split. Similar to the imputation methods, the method in (T.Y. Guo and J.V. Milanovic, 2013) is also exposed to large classification error due to its data replacement mechanism, which can lead to significant degradation in SA accuracy. Another DT-based method is the DT ensemble model proposed in (M. He, *et al.*, 2013). It uses random subspace method to decide the input features for each single DT in the ensemble model, which guarantees that a significant portion of single DTs is available for on-line SA with a high likelihood. In this method, degenerated single DTs are vastly produced for SA under missing-data conditions, which will significantly impair the SA accuracy.

This chapter highlights the missing-data issue in power system SA, and presents four novel data-driven ISs to enhance the on-line and real-time power system SA/SP robustness against the missing-data events. The main feature of these ISs is that they can maintain a high accuracy when the measurement data is only partially available. The developed ISs have been tested on New England 39-bus system, and have demonstrated superior tolerance on the missing-data issues compared to existing approaches.

8.2 Robust Ensemble IS for On-line SA with Missing-data

To improve the on-line SA robustness against missing-data, this section develops a robust ensemble IS that can sustain on-line SA accuracy under PMU missing conditions. This IS consists of a RVFL ensemble model trained with observability-constrained feature subsets to ensure its availability under any PMU missing scenario. The proposed observability-constrained feature subsets are sampled from the full feature set depending on the observability of different PMU combinations, and the parameters involved in RVFL training and ensemble output aggregation are tuned through evolutionary computing. This section first introduces the PMU observability concept and then presents the technical details of the developed robust ensemble IS.

8.2.1 PMU Observability

PMU is a time-synchronized measurement device that is able to measure the bus voltage phasor and incident line current phasors in real-time. Based on the phasor measurements, the voltage phasor of the neighboring buses can be calculated using the known line parameters. Therefore, the basic observability of a single PMU includes the voltage on the installed bus, the power flow on all incident lines, and the voltage on all neighboring buses (B. Gou, 2008; S. Chakrabarti and E. Kyriakides, 2008; S. Chakrabarti, *et al.*, 2008; F. Aminifar, *et al.*, 2009). Moreover, in most cases, the effect of zero injection buses (ZIB) is assumed in PMU observability calculation: in a grid section including a ZIB and all its incident buses, if all buses are observable except one, the only unobservable bus remains observable by using Kirchhoff's current law (KCL) (B. Gou, 2008; S. Chakrabarti and E. Kyriakides, 2008; S. Chakrabarti, *et al.*, 2008; F. Aminifar, *et al.*, 2009). When ZIB is considered, the single PMU observability is sometimes expandable on top of its basic observability.

Here is an example to intuitively illustrate the concept of observability. A simple power network is shown in Fig. 8.1(a) where bus 4 and bus 6 are ZIBs. Two PMUs are installed separately at bus 5 and bus 8. With such PMU placement, the observability of PMU1 alone includes buses 4, 5, 6, lines 4-5, 5-6, and the load at bus 5; and the observability of PMU2 alone is buses 2, 7, 8, 9, lines 2-8, 7-8, 8-9 and generation at bus 2. With both PMU1 and PMU2, line 4-9 and 6-7 becomes observable with known voltages on their ending buses. Since buses 4 and 6 are ZIBs, line 1-4 and 3-6 are also observable by applying KCL. Thus buses 1 and 3 also become observable. Similarly, the loads at buses 7, 9, and the generation at buses 1, 3 can be simply derived by applying KCL. Therefore, complete observability is achieved. It is worth noting that some operating features are observable with the measurement from a single PMU, while some others are only observable when both PMUs are available. The least PMUs needed for a feature to be observable are labelled on the network diagram in Fig. 8.1(b). It can be seen that the operating features can be rearranged into three subsets according to their requirements on PMU availability. The features in each subset are constrained by the observability of either a single PMU or a combination of PMUs. In doing so, during various PMU missing events, there will be at least one subset with completely available features. Such observability-constrained feature subset concept is the basis for the developed robust ensemble IS and the key to mitigate the missing-data issue in on-line SA.

8.2.2 Sampling Mechanism of Feature Subsets

In a power system with PMU as the sensing instrument, the measurement of a single PMU is generally constrained to a small section of the grid.

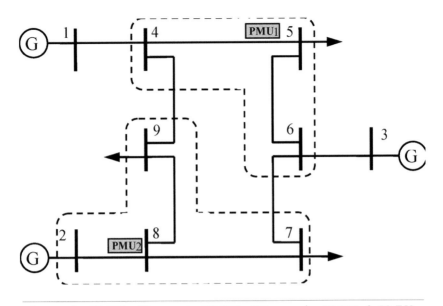

I - require PMU1   II - require PMU2   III - require PMU1 & PMU2

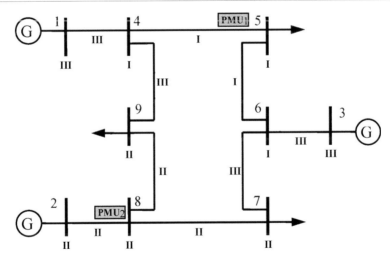

Fig. 8.1: An example of power network with illustrations of (a) the observability of single PMUs in the network, and (b) the required PMUs to observe each feature in the network

By incorporating the measurements from different PMUs, complete observability of the grid can be achieved. However, during PMU missing events, a portion of the operating features cannot be timely obtained and the data-driven models for on-line SA will also become inapplicable. To avoid such a ripple effect, a feasible solution could be to rearrange the whole set of operating features into feature subsets, depending on PMU observability. The aim of doing this is that in case some PMUs are unavailable on-line, there are still feature subsets that are observable by the remaining PMUs and the data-driven models. In this concept, the selected features in each subset are constrained by grid observability in the presence of either a single or multiple PMUs. In other words, the developed feature subsets are *observability-constrained*.

For a power system with N PMUs, there are two extreme ways to sample the feature subsets: one is to compose the observability of each possible PMU combination as a feature subset. In doing so, there will always be complete feature subsets under any PMU missing scenario. However, the number of feature subsets is $2^N - 1$, which can be too large to be computationally practical for a real-world power system with hundreds of PMUs placed. The other way is to take the observability of each single PMU as a feature subset, which leads to the least computation complexity (i.e. N feature subsets). However, since some operating features are observable only in the presence of multiple PMUs, such features will not be excluded from any feature subset, leading to dramatic loss of operating information. Therefore, a new feature sampling method is developed, not only to preserve all the operating features in the system, but also to reduce the number of the sampled subsets.

The sampling mechanism of the feature subsets is illustrated in Fig. 8.2. For a power network with a given placement of N PMUs, the candidate feature set **C** includes the following features: bus voltage magnitude and angle, active and reactive line flow, and net active and reactive power injection to nonzero-injection buses. These candidate features should be collected with complete grid observability. Then, an empty selected feature pool **S** is initialized. It is used for uniting the features that have already been included in any feature subset. The whole sampling process terminates once **S** = **C**, indicating that the provided features in **C** have been comprehensively sampled into feature subsets. In Fig. 8.2, i is the number of available PMUs in the grid and j is the identification number of different PMU combinations. The number of considered PMUs for each feature subset should be as small as possible so that a large number of feature subsets can be available under PMU missing conditions. Therefore, the sampling process starts with $i = 1$ and incrementally increases. For each i, each possible combination of i PMUs is considered and the corresponding observability-constrained feature subset $\mathbf{F}^i_j$ is obtained. If $\mathbf{F}^i_j$ contains any feature that has not been included in **S** ($\mathbf{F}^i_j \subseteq \mathbf{S}$ is not satisfied), $\mathbf{F}^i_j$ will be added to the feature subset pool **P**, and **S** will also be updated by uniting $\mathbf{F}^i_j$; otherwise $\mathbf{F}^i_j$ is considered unnecessary and will be abandoned to reduce the number of sampled subsets. Eventually, **P** consists of all the successfully sampled feature subsets and is delivered as the output.

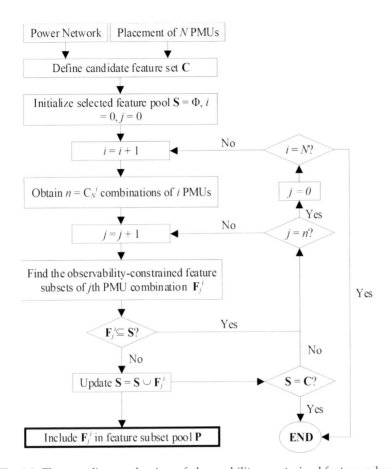

Fig. 8.2: The sampling mechanism of observability-constrained feature subsets

By applying the developed feature sampling method, it can be guaranteed that there is at least one complete feature subset in **P** under any PMU missing condition. Also, all the features in **C** are sampled at least once into the subsets, meaning that the information provided in **C** has been completely preserved in the sampled feature subsets. Moreover, the condition $F^i_j \subseteq S$ is checked in each iteration to avoid excessive computation burden at the subsequent ensemble learning stage. Note that the sampling process can be implemented at PMU planning stage. So its consumed computation time is not relevant to SA computation.

8.2.3 RVFL Training Using Evolutionary Computing

RVFL, as a randomized learning algorithm, has been applied for real-time FIDVR prediction in Section 6.8, and has demonstrated excellent real-time

SP performance. However, in Section 6.8, the number of hidden nodes and the activation function of RVFL are empirically selected, which may not achieve the optimal RVFL performance. Therefore, the RVFLs in the robust ensemble IS are trained through evolutionary computing to pursue the best classification performance of RVFL.

In RVFL training, since the input weights and biases are randomly selected, the output weight vector β becomes the only quantity to be decided. There are two non-iterative methods to obtain β (Y. Ren, *et al.*, 2016; L. Zhang and P.N. Suganthan, 2016): one is Moore-Penrose generalized inverse as used in Section 6.8, $\beta = \mathbf{H}^{\dagger}\mathbf{T}$, and the other method is ridge regression, $\beta = (\mathbf{H}^{T}\mathbf{H} + \lambda\mathbf{I})^{-1}\mathbf{T}$, where λ is a regularization parameter. To choose the best calculation method for SA problem, a test is conducted on New England 39-bus system. The training and testing datasets respectively consist of 4,000 and 2,000 OPs which are sampled by Monte-Carlo technique. The RVFL classifier is tested on eight different contingencies. The testing performance of these two methods is shown in Fig. 8.3(a) where Moore-Penrose generalized inverse outperforms ridge regression. Besides, for activation function selection, four candidate functions are tested, and the testing result is presented in Fig. 8.3 (b) where *sigmoid* function demonstrates the highest accuracy. Therefore, Moore-Penrose generalized inverse is used to compute the output weights and *sigmoid* is selected as the activation function.

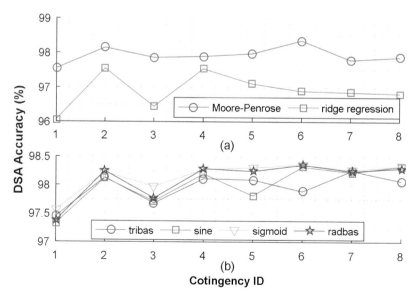

Fig. 8.3: Compare the RVFL performance on SA problem: (a) Moore-Penrose inverse v.s. ridge regression for output weights calculation; and (b) with different activation functions

According to (L. Zhang and P.N. Suganthan, 2016), to pursue the best performance of RVFL classifiers, the randomization range of input weights should be optimized. Instead of directly determining the randomization range, a quantile scaling method proposed in (Y. Ren, *et al.*, 2016) is employed here to re-distribute the randomly weighted input data of RVFL so that the majority of them fall into the non-linear region of the activation function. The quantile scaling algorithm is as follows (Y. Ren, *et al.*, 2016):

(1) For each hidden node, form an input vector $\mathbf{E}_j = \left\{ \sum_{m=1}^{M} w_{m,j} x_m^n + b_{i,j} \right.$

$\left. \right\}$, where N is the number of training instances, M is the number of training features, J is the number of hidden nodes, and x_m^n is the mth feature value of the nth instance. $\mathbf{E}_j$ should be an $N \times 1$ vector.

(2) Obtain the r and $(1 - r)$ quantiles of the vector $\mathbf{E}_j$, Q_r and Q_{1-r}.

(3) Obtain the r and $(1 - r)$ quantiles of the *logsig* function, S_r and S_{1-r}.

(4) Scale $\mathbf{E}_j$ according to   .

In this quantile scaling method, a new parameter r is defined and needs to be tuned. Besides, similar to other SLFNs, the number of hidden nodes J should also be tuned for RVFL. This chapter uses evolutionary computing to optimize J and r in single RVFL training to pursue the best RVFL performance. The training proceeds in an iterative way as illustrated in Fig. 8.4. The input weights and biases of RVFL are randomized in the first iteration and are kept constant in the subsequent process. In each iteration, the populated RVFLs are trained by 80 per cent of the instances randomly sampled, and the rest 20 per cent are used for testing. Eventually, the RVFL with the highest fitness value is preserved as the final RVFL classifier.

8.2.4 Implementation of the Robust Ensemble IS

In the literature and Chapter 4, ensemble learning has been extensively applied to power system on-line SA for higher accuracy and better robustness (I. Kamwa, *et al.*, 2010; N. Amjady and F. Majedi, 2007; M. He, *et al.*, 2013; N. Amjady and S.A. Banihashemi, 2010). In most of the existing ensemble models, the training datasets for each single learning unit is randomly sampled from the database. Although such randomness can improve the learning generality of the ensemble model, its robustness to various missing-data scenarios is uncertain, which is not favorable for practical SA application. Therefore, to maintain the SA robustness against any possible PMU missing event, we incorporate the observability-constrained feature subsets and RVFL ensemble model in the developed

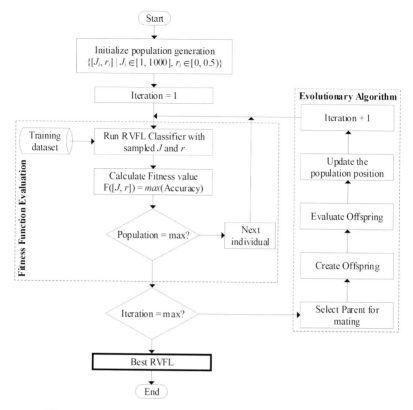

Fig. 8.4: RVFL training process using evolutionary computing

robust ensemble IS. The basic idea of the robust ensemble IS is to use the observability-constrained feature subsets to decide the training features of individual RVFL classifiers in the ensemble model, so that only the available RVFL classifiers under PMU missing condition respond to on-line SA request. As verified in Section 8.2.2, there will be at least one available observability-constrained feature subset under any PMU missing condition. So the ensemble model trained using such feature subsets will also be available. The full computation schematic of the robust ensemble IS is shown in Fig. 8.5, including (a) ensemble training scheme, (b) aggregation validation scheme, and (c) on-line SA application scheme. These three schemes will be elaborated individually in the sequel.

8.2.4.1 *Ensemble Training Scheme*

In the ensemble training scheme in Fig. 8.5(a), if F observability-constrained feature subsets are sampled, the database is thereby rearranged into F training datasets. Each training dataset is responsible to train one RVFL

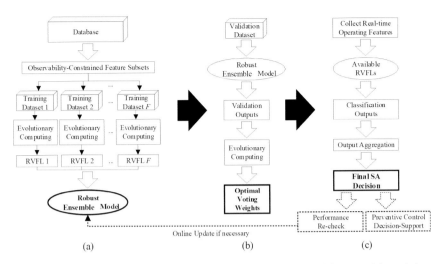

Fig. 8.5: Computation schematic of the robust ensemble IS: (a) ensemble training scheme; (b) aggregation validation scheme; and (c) on-line SA application scheme

classifier using evolutionary computing. The ensemble of the *F* well-trained RVFL classifiers serves as a robust ensemble model.

In such a training scheme, a database is firstly needed as the basis to train individual RVFL classifiers. The OPs of instances in the database are either collected from historical system operation data or empirically sampled from the practical operating space. The security condition of each instance is obtained by running TDS. Therefore, with known security conditions, the instances in the database contain prior knowledge to the SA problem. In data rearrangement, all the instances in original database are mapped into each training dataset, whereas the training features in the training datasets are indicated by the observability-constrained feature subsets.

8.2.4.2 Aggregation Validation Scheme

Following the ensemble training process, the next step is to aggregate the outputs from single learning units to make the final SA decision. In the literature, the final ensemble results are mostly decided based on majority voting (L.K. Hansen and I. Salamon, 1990; Y. Ren, *et al.*, 2016; Z.H. Zhon, *et al.*, 2002), which disregards the uneven training qualities and the different performances of the single learning units. In the ensemble training scheme, since RVFL is an RLA, and each RVFL is trained with a different number of features, the training quality among the single RVFLs can be highly volatile. Thus the performance of the robust ensemble model will be degraded if majority voting is adopted.

To improve the SA performance, a weighted voting strategy is used to aggregate the outputs from single RVFLs. The mechanism of weighted voting is to assign unequal voting weights to individual learning units and the final decision is made, based on the weighted sum of the outputs from single learning units. In this case, the voting weights of every single RVFL should be optimized before they are applied on-line, which is implemented under the aggregation validation scheme shown in Fig. 8.5(b). In such a scheme, the validation dataset is prepared similarly with the database, but in a smaller size. Then, the well-trained, robust ensemble model is applied to the validation dataset and the voting weights are optimized, based on the validation outputs. Here, evolutionary computing is also employed to search for the optimal voting weights for the optimization problem presented below:

$$\underset{\mathbf{w}}{\text{Min}}\, e(\mathbf{w}) = \frac{\sum_{i=1}^{N_v}\left|t_i - \text{Classify}(\mathbf{w}\cdot\hat{\mathbf{t}}_i)\right|}{2\times N_v} \tag{8.1}$$

$$\text{s.t.}\ \sum_{j=1}^{F} w_j = 1 \tag{8.2}$$

$$0 < w_j < 1, j = 1, 2, \ldots, F \tag{8.3}$$

where $\mathbf{w}$ consists of the voting weights of the F RVFL classifiers in the ensemble; the objective $e(\mathbf{w})$ represents the classification error on the validation instances; t_i is the target security status of the ith instance in the validation dataset; and $\hat{\mathbf{t}}_i$ is the vector consisting of the single RVFL classification outputs. The function Classify($\cdot$) classifies the numerical results into either 1 (secure) or –1 (insecure) class based on the following rule:

$$\textbf{If } y > 0, \text{Classify } (y) = 1 \tag{8.4}$$

$$\textbf{Else If } y \leq 0,\ \text{Classify } (y) = -1 \tag{8.5}$$

8.2.4.3 On-line SA Application Scheme

In Fig. 8.5(c), based on the operating features collected in real-time, the robust ensemble model is applied to make on-line SA decision. If there is no PMU missing, full PMU measurements are used. In case of PMU missing events, only available RVFL classifiers are applicable. Based on the observability-constrained feature subsets, there should be at least one available RVFL classifier under any possible PMU missing situation, demonstrating the strong robustness of the developed robust ensemble IS. The weighted voting strategy is then applied to make the final SA decision, which will provide preventive control decision-support.

8.2.5 Achieved Robustness against Missing-data Events

It has been explained that the developed robust ensemble IS remains available for on-line SA under any PMU missing scenario, that is, it has great robustness against PMU missing events at on-line stage. However, besides PMU missing events, it is also necessary to discuss the robustness of the developed IS against other missing-data events, such as communication interrupt, data congestion, and PDC failure.

Communication interrupt can temporarily stop the data transmission from one or several PMUs to the control center. Thus its impact on SA is equivalent to the direct missing of the terminal PMUs with the corrupted communication link. In this sense, the developed robust ensemble IS is capable of dealing with communication interrupt since the IS will remain available to provide on-line SA service, regardless of the missing of the terminal PMUs.

Under data congestion condition, there will be a time-delay between normal data and congested data to be received at the control center, that is, the data synchronism will be lost. In this case, only the time-synchronized data should be utilized for on-line SA, so that the PMUs providing the asynchronous data can be regarded as missing from WAMS. Since the data congestion can be seen as a PMU missing event, the developed robust ensemble IS should be able to ride through a data congestion event in the same manner as a PMU missing event.

A PDC in WAMS is to concentrate the PMU measurements in an area and submit them to the control center. The failure of a PDC can lead to the loss of measurement data in an area, and all its downstream PMUs can be regarded as missing. In this PMU missing situation, the SA process will not be bordered by using the robust ensemble IS.

Therefore, the above missing-data events can all be equivalently seen as PMU missing events, which can be promptly handled by the robust ensemble IS.

8.3 Numerical Test for the Robust Ensemble IS

The robust ensemble IS is tested on New England 39-bus system with the generator on bus 37 being replaced by a wind farm with the same capacity (*see* Fig. 4.22), which aims to reflect the impact of RES. The generated wind power varies between zero and its rated value, and the load-demand levels are assumed to vary between 0.6 and 1.2 of their nominal values.

8.3.1 Database Generation

The OPs in the database are obtained based on the Monte-Carlo technique that randomly samples wind power and loads within their predefined ranges. Then, the other candidate operating features are obtained by running OPF in order to minimize the generation cost. The contingencies considered in the test are the three phase faults on buses with inter-area corridor trip and cleared 0.25 second after their occurrences. Those $N - 1$ contingencies are selected because the disconnection between areas can lead to large disturbances in normal grid operation, resulting in a high probability of insecurity. The security status subject to the selected contingencies is obtained by running TDS using TSAT 16.0 package (D.P. Tools, 2013). Eventually, 7,127 OPs with their security status are simulated. The numbers of secure and insecure instances induced by each contingency are presented in Table 8.1. In the test, 60 per cent of the instances are randomly selected to build the database, 20 per cent to build the validation dataset, and the remaining 20 per cent to build a testing dataset to test the on-line SA performance of the developed model.

Table 8.1: Selected contingencies for robust ensemble IS

Contingency ID	1	2	3	4
Fault setting	Fault bus 3 trip 3-4	Fault bus 4 trip 3-4	Fault bus 14 trip 14-15	Fault bus 15 trip 14-15
No. of secure instances	4350	4343	4327	4440
No. of insecure instances	2710	2717	2733	2620
Contingency ID	**5**	**6**	**7**	**8**
Fault setting	Fault bus 15 trip 15-16	Fault bus 16 trip 15-16	Fault bus 16 trip 16-17	Fault bus 17 trip 16-17
No. of secure instances	4430	4448	4471	4452
No. of insecure instances	2630	2612	2589	2608

8.3.2 PMU Placement

In PMU placement, although single branch outages (S. Chakrabarti and E. Kyriakides, 2008) and measurement redundancy (N. Amjady and S.A. Banihashemi, 2010) are also important, this test only considers placing the minimum number of PMUs in order to avoid unnecessary costs, while complete observability of the grid can be achieved. According to (S. Chakrabarti and E. Kyriakides, 2008), at least eight PMUs are required to achieve complete observability of New England 39-bus system and there are four possible PMU placement options, which are listed in Table 8.2. It can be seen that the only difference between options I-III is the PMU placed on bus 10, 12, or 13, and the PMU placed on bus 3 is relocated to bus 18 in option IV. By applying the proposed feature sampling method, the number of obtained observability-constrained feature subsets for the four PMU placement options are different as shown in Table 8.2.

Table 8.2: Specifications of PMU placement and the obtained feature subsets

PMU placement options	Located buses	No. of PMUs	No. of feature subsets
I	3, 8, 10, 16, 20, 23, 25, 29	8	18
II	3, 8, 12, 16, 20, 23, 25, 29	8	17
III	3, 8, 13, 16, 20, 23, 25, 29	8	18
IV	8, 13, 16, 18, 20, 23, 25, 29	8	20

8.3.3 Model Training and Aggregation Validation

Based on the observability-constrained feature subsets and the training scheme in Fig. 8.5(a), a robust ensemble model is trained for each contingency. The number of RVFL classifiers in each ensemble should equal the number of feature subsets. Due to the fast learning speed of RVFL and the few parameters to be optimized through evolutionary computing, the training efficiency of every single RVFL is very high. In the test, training one single RVFL classifier spends 15.8 seconds on an average, using a PC with 3.3GHz CPU. Moreover, since the number of RVFLs to be trained in the ensemble is small, i.e. 18, 17, 18, and 20 respectively for PMU placement options I -IV, the efficiency of off-line ensemble training is also very high.

The validation dataset is then applied to optimize the voting weights following the process in Fig. 8.5(b). At on-line stage, if any RVFL in the ensemble becomes unavailable due to PMU missing events, the optimal voting weight assigned to that RVFL will be temporarily disabled. In the test, genetic algorithm is employed for evolutionary computing.

8.3.4 Testing Result

The well-trained robust ensemble model and optimal voting weights are applied to the testing dataset, considering all possible PMU missing scenarios. The more PMUs are missing in the system, the fewer RVFLs in the ensemble remain available. Thus lower SA accuracy should be an inevitable result. The testing results corresponding to PMU placement options I-IV are demonstrated in Fig. 8.6. For a fixed number of missing PMUs indicated in the horizontal axis, the average SA accuracy over different possible PMU missing scenarios is shown. It can be seen from Fig. 8.6 that for most of the contingencies, the degradation of SA accuracy between complete observability and single PMU observability is smaller

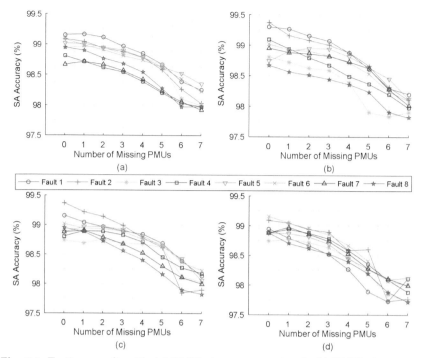

Fig. 8.6: Testing result with (a) PMU placement option I, (b) PMU placement option II, (c) PMU placement option III, and (d) PMU placement option IV (*Note*: The testing result subject to each contingency is shown separately in each subplot)

than one per cent, which can be negligible when verifying the excellent robustness of the developed robust ensemble IS.

8.3.5 Comparative Study

To demonstrate the advantages of the robust ensemble IS over other existing methods, a comparative test is conducted to compare the developed IS with decision tree with surrogate split (DTSS) method (T.Y. Guo and J.V. Milanovic, 2013), which uses surrogate splits to substitute the DT splitting rules that contain missing features. Both methods are applied to the same SA problem, i.e. the same power network, the same database, the same validation dataset, and the same testing set. For DTSS method, the database is used to train the DT and the validation dataset is used to find the best pruning level of the trained DT.

In real-world measurement system, the probabilities of conservative and extreme PMU missing events are very different. For instance, failure of any single PMU in WAMS is commonly experienced, but a simultaneous unavailability of a large portion of PMUs is relatively rare. Taking the above probability concept into consideration, the robustness of the two methods is compared with respect to different PMU availability rates. The PMU availability rate refers to the portion of time for a single PMU to be available in the system. For a given PMU availability rate a, the overall SA accuracy A^{SA} is evaluated as follows:

$$A^{DSA} = \sum_{i=0}^{N-1} C_N^i (1-a)^i a^{N-i} \overline{A}_i \qquad (8.6)$$

where i represents the number of missing PMUs in the system; C_N^i is the number of all possible PMU missing scenarios for i missing PMUs; and $\overline{A}_i$ is the average SA accuracy over all the considered contingencies of all C_N^i PMU missing scenarios.

In the test, PMU availability rate a decreases from 100 per cent to 90 per cent with a step size of one per cent, and a is assumed identical to each PMU in the system. As PMU availability rate a decreases, the probability of PMU missing events becomes higher. So the overall SA accuracy A^{SA} will inevitably decrease. In this case, the decreasing slope of the overall SA accuracy becomes an indicator to compare the robustness of the two methods. In Fig. 8.7, when PMU measurements are fully available (the leftmost points), the SA accuracy of robust classification model achieves around 99 per cent while the accuracy of DTSS model is 98.22 per cent; at 90 per cent PMU availability rate (the rightmost points), the robust classification model can still maintain around 99 per cent accuracy, but the accuracy of DTSS sharply drops to 92.76 per cent, 93.86 per cent, 94.54 per cent and 94.54 per cent respectively for the four PMU placement options.

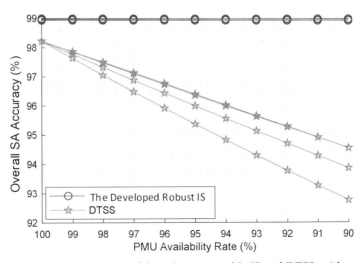

Fig. 8.7: Overall SA accuracy of the robust ensemble IS and DTSS, with respect to different PMU availability rates (*Note*: The SA accuracy using different PMU placement options are plotted separately for each method)

It can be seen that, with decreasing PMU availability rate, the accuracy of the robust ensemble IS sustains while the performance of DTSS degrades dramatically. Therefore, the developed robust ensemble IS demonstrates much higher accuracy as well as better robustness against missing-data events.

8.4 Robust Ensemble IS: To be Mathematically Rigorous

In previous sections, the robust ensemble IS showed a promising performance on on-line SA in the presence of missing-data events. Nevertheless, in robust ensemble IS, the mechanism of sampling the observability-constrained feature subsets is empirically designed, which lacks mathematical rigorousness on its feasibility and optimality. With this concern, this section upgrades the robust ensemble IS by replacing the feature sampling process with a strategically designed PMU clustering process. Trained by the features observed by each PMU cluster, the ensemble model ends up with the minimum number of single classifiers. Moreover, under any PMU missing scenario, this upgraded robust ensemble IS is able to maximize the utilization of different PMUs to maintain the system observability and thereby the SA accuracy. The optimality of this IS is also mathematically proved in this section. This upgraded IS is tested on New England 39-bus system under various PMU placement strategies. The testing results demonstrate its exceptional SA tolerance to missing-data events in measurements systems.

8.4.1 Working Principle

The working principle of the upgraded IS is illustrated in Fig. 8.8. To tackle the missing-data issue in on-line SA, a set of single classifiers is built, each trained by a strategically selected cluster of PMU

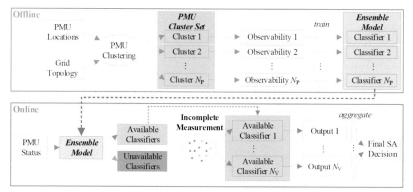

Fig. 8.8: Working principle of the upgraded robust ensemble IS

measurements which covers a region of the observable grid. At the on-line stage, the single classifiers make individual SA decisions based on their own PMU measurements. With complete PMU measurements, the whole set of classifiers ensures full observability of the system. If some PMU measurements are missing, the corresponding classifiers become 'unavailable'. Then the SA decision is only made by the 'available' classifiers which receive complete data inputs. The final SA result is the aggregated output of the 'available' classifiers using the weighted voting method.

By strategically clustering the PMUs, the upgraded robust ensemble IS can guarantee that under any PMU missing scenario, at least one classifier is available and the observable system measurements of remaining PMUs will be fully covered. Thus SA accuracy can be sustained to the maximum extent. Meanwhile, this upgraded IS can minimize the number of PMU clusters and the corresponding single classifiers, so that the computation complexity is minimal.

8.4.2 PMU Clustering Algorithm

In the above upgraded IS, PMU clustering is the key step to achieve its robustness and optimality. A novel PMU clustering algorithm is presented in this section to search for the minimum number of PMU clusters that can fully cover the available observability under any PMU missing condition. The algorithm of PMU clustering is presented in two forms, including a piece of pseudocode in Algorithm 8.1 and a flow diagram in Fig. 8.9.

Algorithm 8.1: PMU Clustering Algorithm

For a power network with N_E electric components and N PMUs:
1: Define a set $\{e_i, i = 1, \ldots, N_E\}$, including all the electric components
2: **for** $i = 1$ **to** N_E, **do**
3:     Initialize $\mathbf{R}_i = \varnothing$, a set of vectors to collect PMU clusters for e_i
4:     **for** $j = 1$ **to** $N - 1$, **do**
5:         Define a set of vectors $\{\mathbf{m}_k, k = 1, \ldots, C_N^j\}$, where $\mathbf{m}_k$ is a vector denoting a combination of j out of N PMUs.
6:         **for** $k = 1$ **to** C_N^j, **do**
7:         **if** $\mathbf{m}_k$ observes e_i & no vector in $\mathbf{R}_i \subseteq \mathbf{m}_k$ ($\mathbf{m}_k$ is non-redundant)
8:         $\mathbf{R}_i = \mathbf{R}_i \cup \mathbf{m}_k$
9:         **end if**
10:         **end for**
11:     **end for**
12: **end for**
13: **Return** the PMU cluster set $\mathbf{P} = \bigcup_{i=1\ldots N_E} \mathbf{R}_i$, $\mathbf{P}$ is a set of vectors

---

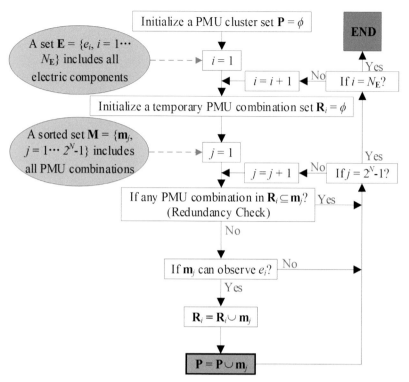

Fig. 8.9: PMU clustering process

In the above algorithm, if a PMU combination **c** observes a component e_i while none of the subsets of **c** can observe e_i, we define **c** as a non-redundant combination for e_i. For each component e_i, all combinations that can observe e_i are iteratively searched and only the non-redundant combinations are collected in $\mathbf{R}_i$. All $\mathbf{R}_i$ are finally united as the PMU cluster set **P**. Here, we call each PMU combination in **P** a cluster and **P** can satisfy the following two conditions under any PMU missing scenario:

F1 *The union of the observability of each complete cluster in* **P** *equals the remaining observability of the grid.*

F2 *Upon* **F1** *getting satisfied, the number of clusters is minimized.*

In on-line SA, **F1** ensures the full coverage of the available PMU observability while **F2** ensures the minimum computation complexity. Besides, the algorithm is unified to any PMU placement, either with or without measurement redundancy and/or ZIB effect.

8.4.3 Mathematical Proof

Mathematical proof of **F1** and **F2** is given below:

F1 proof:

$$\textit{F1 is equivalent to: } \mathbf{E}_1 = \mathbf{E}_2, \ \forall \ \mathbf{d} \in \mathbf{M} \tag{8.6}$$

where

$$\mathbf{E}_1 = O(\mathbf{d}), \ \mathbf{E}_2 = \bigcup_{\mathbf{m}_k \in \mathbf{P}} O(V(\mathbf{m}_k \mid \mathbf{d})) \tag{8.7}$$

where

$$V(\mathbf{m}_k \mid \mathbf{d}) = \begin{cases} \mathbf{m}_k \text{ if } \mathbf{m}_k \subseteq \mathbf{d} \\ \phi \text{ otherwise} \end{cases} \tag{8.8}$$

In (8.6)-(8.8), $O(\cdot)$ maps the observability of a set of PMUs; $\mathbf{d}$ is the set of available PMUs; $\mathbf{M}$ includes all PMU combinations; $\mathbf{m}_k$ is a PMU cluster in $\mathbf{P}$ and the condition $\mathbf{m}_k \subseteq \mathbf{d}$ means $\mathbf{m}_k$ remains complete with the only $\mathbf{d}$ in the system.

$\forall e_i \in \mathbf{E}_1 = O(\mathbf{d})$, at least one non-redundant subset $\mathbf{d}_s \subseteq \mathbf{d}$ satisfies $e_i \in O(V(\mathbf{d}_s \mid \mathbf{d}))$. Since $\mathbf{R}_i$ includes all non-redundant PMU clusters for e_i, $\mathbf{d}_s \in \mathbf{R}_i \subseteq \mathbf{P}$, thus $e_i \in \mathbf{E}_2 \Rightarrow \mathbf{E}_1 \subseteq \mathbf{E}_2$. $\forall e_i \in \mathbf{E}_2$, at least a $\mathbf{m}_s \in \mathbf{P}$ satisfies $e_i \in O(\mathbf{m}_s)$ and $\mathbf{m}_s \subseteq \mathbf{d}$, so $e_i \in O(\mathbf{d}) = \mathbf{E}_1 \Rightarrow \mathbf{E}_2 \subseteq \mathbf{E}_1$. As $\mathbf{E}_1 \subseteq \mathbf{E}_2$ and $\mathbf{E}_2 \subseteq \mathbf{E}_1$, $\mathbf{E}_1 = \mathbf{E}_2 \Rightarrow$ *F1* is true.

F2 proof:

We make a hypothesis H: There is a PMU cluster $\mathbf{m}_a$ that can be removed from $\mathbf{P}$ and $\mathbf{P} \setminus \mathbf{m}_a$ still satisfies (8.6).

Let $\mathbf{d} = \mathbf{m}_a$, $e_b \in \mathbf{E}_1 = O(\mathbf{m}_a)$, and $\mathbf{m}_a \in \mathbf{R}_b$. As the clusters in $\mathbf{R}_b$ are non-redundant, all the clusters in $\mathbf{R}_b \setminus \mathbf{m}_a$ include at least one PMU that is not in $\mathbf{m}_a$, so $\mathbf{m}_{k1} \not\subset \mathbf{d}, \forall \ \mathbf{m}_{k1} \in \mathbf{R}_b \setminus \mathbf{m}_a$. As $\mathbf{R}_b$ includes all clusters observing e_b, $\mathbf{P} \setminus \mathbf{R}_b$ cannot observe e_b, thus

$$\begin{cases} O(V(\mathbf{m}_{k1} \mid \mathbf{m}_a)) = \phi, \ \forall \mathbf{m}_{k1} \in \mathbf{R}_b \setminus \mathbf{m}_a \\ e_b \notin O(V(\mathbf{m}_{k2} \mid \mathbf{m}_a)), \forall \mathbf{m}_{k2} \in \mathbf{P} \setminus \mathbf{R}_b \end{cases} \Rightarrow e_b \notin O(V(\mathbf{m}_k \mid \mathbf{m}_a)),$$

$$\forall \mathbf{m}_k \in \mathbf{P} \setminus \mathbf{m}_a \ \Rightarrow e_b \notin \mathbf{E}_2 \Rightarrow \mathbf{E}_1 \neq \mathbf{E}_2$$

Thus, H fails $\Rightarrow$ *F2* is true.

8.4.4 Numerical Test

The upgraded robust ensemble IS is applied on New England 39-bus system to numerically test its SA performance. In the test, we apply two benchmark PMU placement strategies reported in (S. Chakrabarti and E. Kyriakides, 2008) with the minimum number of PMUs required to achieve global observability. The two PMU placement strategies, as well as their corresponding PMU clusters, are shown and listed in Fig. 8.10. For PMU placement 1, the ZIB effect is assumed in observability calculation, and the eight PMUs are rearranged into 19 clusters by the PMU clustering

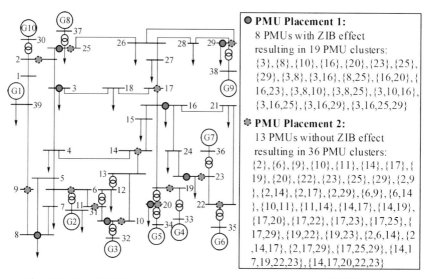

PMU Placement 1:
8 PMUs with ZIB effect
resulting in 19 PMU clusters:
{3},{8},{10},{16},{20},{23},{25},
{29},{3,8},{3,16},{8,25},{16,20},{
16,23},{3,8,10},{3,8,25},{3,10,16},
{3,16,25},{3,16,29},{3,16,25,29}

PMU Placement 2:
13 PMUs without ZIB effect
resulting in 36 PMU clusters:
{2},{6},{9},{10},{11},{14},{17},{
19},{20},{22},{23},{25},{29},{2,9
},{2,14},{2,17},{2,29},{6,9},{6,14
},{10,11},{11,14},{14,17},{14,19},
{17,20},{17,22},{17,23},{17,25},{
17,29},{19,22},{19,23},{2,6,14},{2
,14,17},{2,17,29},{17,25,29},{14,1
7,19,22,23},{14,17,20,22,23}

Fig. 8.10: Two PMU placement strategies and the composed PMU clusters

algorithm. In contrast, for PMU placement 2, the ZIB effect is not considered, and then the 13 PMUs are rearranged into 39 clusters.

In the test, 10 three-phase faults at inter-area corridors are studied, and 5,300 different OPs are obtained and simulated to construct the stability database. Eighty per cent and 20 per cent of these instances are respectively used for training and testing purposes. The single classifier is based on ELM algorithm (G.B. Huang, *et al.*, 2006) and their outputs are aggregated, using weighted voting. The upgraded robust ensemble IS is compared with DTSS method (T.Y. Guo and J.V. Milanovic, 2013) in the test. The testing results are shown in Fig. 8.11 where SA accuracy is the percentage of the testing instances that are correctly classified. Each plotted point represents the average SA accuracy over all the scenarios for each identical number of missing PMUs. According to Fig. 8.11, with increased missing data, DTSS dramatically drops in accuracy. By contrast, the upgraded robust ensemble model is not significantly affected, and can maintain a high SA accuracy even with only one available single classifier. To further evaluate the robustness of SA against PMU missing events, a new SA robustness index, *RI*, is proposed as follows:

$$RI = \frac{1}{N-1}\sum_{i=1}^{N-1} r_i, \text{ where } r_i = \frac{A_i}{A_0} \times 100\% \qquad (8.9)$$

where A_i is the average SA accuracy with i PMUs missing and A_0 is the SA accuracy with complete measurement. So r_i is the percentage of the sustained SA accuracy. Then *RI* is the average r_i over all the PMU missing scenarios. A larger *RI* value indicates higher SA robustness against

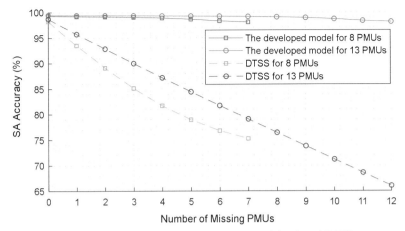

Fig. 8.11: SA testing results of robust ensemble IS and DTSS

missing-data events. The SA robustness indices of the two methods are compared in Table 8.3 where the upgraded robust ensemble IS reports significantly higher SA robustness than DTSS.

Table 8.3: SA robustness index of upgraded robust ensemble IS and DTSS

SA method	PMU placement 1	PMU placement 2
Upgraded Robust Ensemble IS	99.5%	99.6%
DTSS	84.4%	81.7%

8.5 Generative Adversarial Networks for On-line SA with Missing-data

In the previous section, a robust ensemble IS is designed to strategically collect observability-constrained PMU clusters as training datasets. Each of such datasets is further utilized to train single SA classifiers. Under any PMUs missing conditions, the ensemble of available classifiers could achieve maximum system observability with least single classifiers, thus maintaining a high accuracy and computational efficiency. However, it is clear that the success of this IS fully depends on the PMU observability and network topologies. Once they are changed or the IS cannot expect the detailed placement of missing data in advance, the SA model may become ineffective and has to be updated.

To further address the missing-data problem, this paper proposes a fully data-driven IS based on an emerging deep-learning technique, called 'generative adversarial network' (GAN) (I. Goodfellow, *et al.*, 2014; D. Pathak, *et al.*, 2016; Y. Chen, *et al.*, 2018). The principle is to develop a

GAN model to directly and accurately fill up the missing data without depending on PMU observability and network topologies. Thus, the method is more generalized and extensible. Besides, any machine learning algorithm can be used for the SA model. Specifically, this new IS combines a GAN model and a SA model. By constantly updating and fixing GAN model, it can collectively provide a complete data set against the missing data. The SA model can be a classifier or predictor using any effective machine learning algorithm in an ensemble form, in this section as ELM and random vector functional link networks RVFL, to obtain the more diversified machine learning outcome.

Moreover, a high-performance stochastic optimization gradient descent algorithm Adam (D.P. Kingma and J. Ba, 2014) is applied in this section for weight updates during the GAN model training process, to make the training process more accurate and faster. The main contributions of the new IS are as follows:

(1) A GAN model is developed to generate the missing data without depending on PMU observability and network topologies. Adam algorithm is used to pursue the highest assessment accuracy and efficiency.
(2) A hybridized random learning model is developed for SA. Multiple randomized learning algorithms are ensembled to improve the learning diversity. Their aggregate output tends to be more accurate and more robust than the performance of using a single learning algorithm.

The developed IS has been tested on New England 10-machine 39-bus system and demonstrated higher accuracy compared to the other state-of-the-art methods. By offering the system operators more flexibility in manipulating the assessment performance, the developed IS can accommodate different PMU missing conditions.

The nomenclature for this section is given in Table 8.4.

Table 8.4: Nomenclature

Notation	Description	Notation	Description
G	Generator	θ_g	Generator weights
D	Discriminator	θ_d	Discriminator weights
$V(\cdot)$	Value function	x	True dataset
p_g	True sample distribution	z	Input vector
x	Single sample	r	GAN learning rate
m	Number of ELM or RVFL	φ	Clipping parameter
n_{discri}	Number of discriminator iterations	$\mathbf{w}$	Weight
n_{gene}	Number of generator iterations	$\mathbf{b}$	Biases
ρ_1, ρ_2	Exponential decay rates	ε	Prevent parameters

8.5.1 Generative Adversarial Network

As an emerging unsupervised machine learning algorithm, GAN can directly generate realistic features based only on the training data set without the need to fit an existing explicit model (I. Goodfellow, *et al.*, 2014). GAN is implemented with two deep neural networks, called *generator* and *discriminator*, which contest with each other in a zero-sum game framework.

Specifically, the generator is to produce data samples that follow the distribution of the historical training data, while the discriminator is to distinguish between the generated data samples and the true historical data. By training the generator and the discriminator iteratively, GAN can achieve an equilibrium that the discriminator can no longer distinguish between the generated and historical data. This means that the generated data is aligned with the distribution of the historical data. For the missing PMU data problem, GAN is to generate data that can accurately replace the missing data.

Generator: Denotes the true sample distribution of the observation features as p_g over dataset x. We define prior input noise vector z under a given distribution $z \sim p_z(z)$ which can be easily sampled (e.g. incomplete PMU measurements). The objective is to find a function G following $p_g(x)$ as $G(z, \theta_g)$ after transformation, where G is a differentiable function represented by a multilayer perceptron.

Discriminator: Denotes a second multilayer perceptron $D(x, \theta_d)$ that outputs a single scalar, where $D(x)$ represents the probability that x came from the generated data rather than p_g, and thus to maximize the probability of the correct label between $\mathbb{E}[D(x)]$ (real data) and $\mathbb{E}[D(G(z))]$ (generated data).

Here, θ_g and θ_d denote the weights of the generator and discriminator neural networks, respectively.

The structure of GAN is illustrated in Fig. 8.12, and its training process is detailed in Algorithm 8.2.

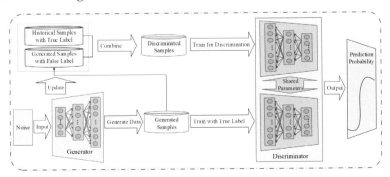

Fig. 8.12: Structure of the GAN

Algorithm 8.2: GAN Training Process

Input: Learning rate r, minibatch size m, clipping parameter φ, number of iterations for discriminator n_{discri} and generator n_{gener}

Output: Complete data $O_{gener}(x)$ filled by generated data $O_{gener}(x)$

Initialize: Initial generator weight θ_g and discriminator θ_d

 for number of training iterations **do**

  for $i = 0, \ldots, n_{discri}$ **do**

   # Update discriminator parameter, fix generator parameter

   Sample minibatch of m examples from historical data:

$$\left\{x^{(t)}\right\}_{t=1}^{m} \sim p_{\text{data}}(x)$$

   Sample minibatch of m noises from prior noise:

$$\left\{z^{(t)}\right\}_{t=1}^{m} \sim p_g(z)$$

   Update discriminator networks by descending its gradient:

$$g_{\theta_d} \leftarrow \nabla_{\theta_d} \frac{1}{m} \sum\nolimits_{t=1}^{m}\left[\log D(x^{(t)}) + \log\left(1 - D\left(G(z^{(t)})\right)\right)\right]$$

  end for

  for $j = 0, \ldots, n_{gener}$ **do**

   # Update generator parameter, fix discriminator parameter

   Sample minibatch of m noises from prior noise:

$$\left\{z^{(t)}\right\}_{t=1}^{m} \sim p_g(z)$$

   Update generator networks by descending its gradient:

$$g_{\theta_g} \leftarrow \nabla_{\theta_g} \frac{1}{m} \sum\nolimits_{t=1}^{m}\left[\log\left(1 - D\left(G(z^{(t)})\right)\right)\right]$$

  end for

 end for

The gradient-based update can use any standard gradient-based learning rule. In this model, Adam algorithm is applied here.

The GAN training can be divided into two steps – updating discriminator with fixed generator parameter and updating generator with fixed discriminator parameter. As the objectives for generator and discriminator defined above, GAN needs to update neural networks' weights based on the loss function. The loss functions for discriminator and generator are respectively formulated as:

$$\max_{D} V(D,G) = \mathbb{E}_{x \sim p_{data}(x)} \Big[\log D(x) \Big] + \mathbb{E}_{z \sim p_z(z)} \Big[\log \big(1 - D(G(z))\big) \Big] \quad (8.10)$$

$$\max_{G} V(D,G) = \mathbb{E}_{z \sim p_z(z)} \Big[\log \big(D(G(z))\big) \Big] \quad (8.11)$$

where (8.11) is equal to (8.12) as follows:

$$\max_{G} V(D,G) = \mathbb{E}_{z \sim p_z(z)} \Big[\log \big(1 - D(G(z))\big) \Big] \quad (8.12)$$

Then we combine (8.10) with (8.12) to formulate a two-player iterative minimax game with value function $V(G, D)$:

$$\min_{G} \max_{D} V(G,D) = \mathbb{E}_{x \sim p_{data}(x)} \Big[\log D(x) \Big] + \mathbb{E}_{z \sim p_z(z)} \Big[\log \big(1 - D(G(z))\big) \Big] \quad (8.13)$$

This objective function can produce the same fixed point of dynamic of generator G and discriminator D, and provides much stronger gradients in the early learning process. When G is less, D can reject samples with high confidence owing to their obvious difference from the training data. In this case, log(1-$D(G(z))$) saturates. Rather than training G to minimize log(1-$D(G(z))$), we can train G to maximize log $D(G(z))$.

We consider the optimal discriminator D for any given generator G. The value function $V(G, D)$ can reach the global maximum optimization when satisfying two conditions, *F1* and *F2*, as follows:

F1. For G fixed, the optimal discriminator D is:

$$(8.14)$$

F2. The global minimum of is *achieved if and only if:*

$$p_g(x) = p_{data}(x) \quad (8.15)$$

The proof is provided in (I. Goodfellow, *et al.*, 2014).

As mentioned before, $D(x, \theta_d)$ and $G(z, \theta_g)$ are completely different functions. Thus any standard gradient-based learning rule for these two networks can be applied to optimize their performances.

8.5.2 Adam Optimization Algorithm

The *Adam* optimization algorithm is an extension of stochastic gradient descent that computes adaptive learning rates for each parameter (D.P. Kingma and J. Ba, 2014). *Adam* optimization algorithm combines the best properties of the *AdaGrad* and *RMSProp* algorithms to handle sparse gradients on noisy problems. In addition to storing an exponentially decaying average of past squared gradients *RMSprop*, Adam also keeps counteracting these biases by computing bias-corrected first and second moment estimates. Instead of adapting the parameter learning rates, based on the average first moment (the mean) as in *RMSProp*, Adam also makes use of the average of the second moments of the gradients (the uncentered variance). The work process is summarized in Algorithm 8.3.

Algorithm 8.3: *Adam* Optimization Algorithm

Input: Learning rate α, exponential decay rates for the momentum ρ_1 and ρ_2, stochastic objective function $f(\theta)$ with parameter θ, small number to prevent any division by zero ε.
Output: Resulting parameter θ_t.
Initialize: Initial parameter θ_0.
    $m_0 \leftarrow 0$ and $v_0 \leftarrow 0$ # 1st and 2nd moment vector
    $t \leftarrow 0$ # timestep
    while θ_t not converged **do**
        $t \leftarrow t + 1$
        # Calculate gradient at timestep t
        $g_t \leftarrow \nabla_\theta f_t(\theta_{t-1})$
        # Updata biased first and second moment estimate

        # Compute bias-corrected first and second moment estimate

        # Updata parameter θ_t
        $\theta_t \leftarrow \theta_{t-1} + \alpha \cdot \hat{m}_t / (\sqrt{\hat{v}_t} + \varepsilon)$
    end while

8.5.3 IS Implementation

The whole framework of the GAN-based IS can be divided into two stages, including off-line training stage and on-line application stage, which are illustrated in Fig. 8.13.

- **Off-line Training Stage**

  At the off-line training stage, the SA database is used to train a GAN model and a SA model. GAN consists of two deep neural networks – the *generator* and the *discriminator*. By constantly updating and fixing each other, the generator can generate the data which can fill up

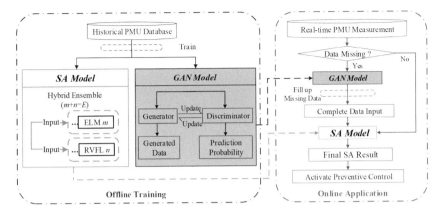

Fig. 8.13: Framework of the GAN-based IS

incomplete PMU data. The discriminator can calculate the prediction probability of corresponding generated data. After that, they can collectively provide an accurate complete data set against missing data. A high-performance stochastic optimization gradient descent algorithm *Adam* is applied for weights updating in both discriminator and generator neural networks. Clipping parameter is utilized to constrain $D(x, \theta_d)$ to satisfy certain practical conditions as well as prevent gradient explosion. The SA model is the classifier based on hybrid randomized ensemble model of ELM and RVFL, as presented in Section 6.6.2.

- **On-line Application Stage**

  At on-line stage, the real-time PMU measurements are the input features. The IS first detects whether the PMU data is complete. If the data is complete, they will be directly imported into the SA model; otherwise, the incomplete data will be fed to the GAN, and GAN will fill up the missing data. After that, the filled, complete data will be imported to the SA model and the classification decision on system's dynamic security status is made by the SA model. If the system is insecure, preventive control actions, such as generation rescheduling, should be activated.

8.6 Numerical Test on GAN-based IS

The developed GAN-based IS is tested on New England 10-machine 39-bus system (*see* Fig. 8.10), where two benchmark PMU placement strategies, installed with the minimum number of PMUs required to realize global observability. The numerical simulation is conducted on a 64-bit computer with an Intel Core i7 CPU of 2.8-GHz and 16-GB RAM. The developed IS is implemented in the MATLAB platform.

8.6.1 Database Generation

Simulations are performed on New England 10-machine 39-bus system to test the developed IS. The OPs are generated using the Monte Carlo method. Specifically, given the forecasted load demand level for each bus, Monte Carlo simulation is run to sample uncertain power variations. Then, for each uncertain power demand scenario, the power output of the generators is determined by OPF calculation. In doing so, a large number of OPs will be obtained. Then, given the contingency set, TDS is performed to evaluate the dynamic security label for each OP. The contingencies considered in the study are the three-phase faults with inter-area corridor trip and cleared 0.25 s after their occurrences. 10 typical three-phase faults at inter-area corridors are simulated on TSAT. Note that any contingencies can be considered depending on the practical needs. Eventually, 5,043 OPs with their security conditions are obtained, and the number of secure instances and insecure instances under each contingency are shown in Table 8.5. Thus, there are a total of 50,430 instances. In the subsequent work, 80 per cent of the instances were randomly selected as for both SA model and GAN model training, and the remaining 20 per cent instances form the SA model testing dataset.

Table 8.5: Contingency set

Contingency ID	1	2	3	4	5
Fault setting	Fault bus 1, trip 1-39	Fault bus 39, trip 1-39	Fault bus 3 trip 3-4	Fault bus 4 trip 3-4	Fault bus 14 trip 14-15
No. of secure instances	3257	3075	3417	3326	3419
No. of insecure instances	1786	1968	1626	1717	1624

Contingency ID	6	7	8	9	10
Fault setting	Fault bus 15 trip 14-15	Fault bus 15 trip 15-16	Fault bus 16 trip 15-16	Fault bus 16 trip 16-17	Fault bus 17 trip 16-17
No. of secure instances	3462	3394	3437	3320	3282
No. of insecure instances	1581	1649	1606	1723	1761

8.6.2 Parameter Selection

(1) *For SA model*:
  - *Quantity of ELM and RVFL in an Ensemble E*: In ensemble learning scheme, the generalization error will decrease as quantity of single learner increases. In this test, the total number is set as 200 with 100 ELMs and 100 RVFLs.
  - *Random Parameters in Ensemble Learning*: For ELM, activation function is chosen as Sigmoid function and the number of hidden nodes of ELM is randomly selected from [200, 400]; for RVFL, the activation function is Sigmoid function and the optimal hidden nodes range is [100,300]. This could also verify that the direct link of RVFL could lead to a thinner model. Sigmoid function is chosen as valid activation hidden optimal range of ELM and RVFL classifiers. And the common optimal hidden node range for ELM and RVFL algorithms is [200, 300].
  - *Quantity of Training Instances and Testing Instances*: The quantity of instances which are selected to train ELMs and RVFLs determines the whole performance of the robustness. In this case study, training instance and testing instance are chosen to be 4,034 and 1,009, respectively.
(2) *For GAN model*:
  - *Learning Rate* α: Learning rate is a hyper-parameter that controls how much we are adjusting the weights of our network with respect to the loss gradient. The value of the learning rate α is decided to be 0.001.
  - *Exponential Decay Rates for the Momentum* ρ_1 *and* ρ_2: In this test, exponential decay rates for the momentum ρ_1 and ρ_2 are set as 0.9 and 0.999, respectively.
  - *Number to Prevent any Division by zero* ε: In this test, ε is very small so as to prevent any division and set as 10^{-8}.
  - *Quantity of Training Instances*: In this test, training instance for GAN model is same as SA model and is chosen to be 4,034.

8.6.3 Testing Results

The developed GAN-based IS is compared with three state-of-the-arts: the updated robust ensemble IS in Section 8.4, DTSS (T.Y. Guo and J.V.Milanovic, 2013), and feature estimation (Q. Li, *et al.*, 2019). Note that all the results are the average value of the 10 test runs for the 10 contingencies.

(1) *Performance on Filling up Missing Data*: Firstly, the GAN's performance on filling up the missing data is tested. The error is evaluated by MAPE (S.S. Armstrong and F. Collopy, 1992).

The results in Table 8.6 are the MAPE for two PMU placement options under all missing PMU scenarios. It can be seen that the prediction of missing data is very close to the real value, which demonstrates that the generated data by the GAN model can accurately fill up the missing data. Moreover, the average computation time are 1.97 ms and 2.24 ms for one instance under two different PMU placement locations respectively. These are negligible for on-line SA application.

Table 8.6: GAN filling missing data accuracy

PMU options	MAPE under different missing PMU numbers (%)											
	1	2	3	4	5	6	7	8	9	10	11	12
I	1.29	1.55	1.71	1.98	2.11	2.27	2.58					
II	1.33	1.42	1.56	1.73	1.79	1.84	2.01	2.06	2.25	2.28	2.55	2.86

(2) *Classification Accuracy Comparison*: The GAN-based IS is compared with two existing methods, the updated robust ensemble IS in Section 8.4 and DTSS (T.Y. Guo and J.V.Milanovic, 2013). The testing results corresponding to PMU placement options I-II are demonstrated in Figs. 8.14 and 8.15, where SA accuracy is the percentage of the testing instances that are correctly classified. Each point represents the average SA accuracy under each identical number of missing random PMUs data. Figures 8.14(a) and 8.15(a) show the average SA accuracy under the 10 selected three-phase faults, Figs. 8.14(b)-(d) and 8.15(b)-(d) show the SA performance on three randomly selected faults, that are fault 1, 5 and 10, under the PMU placement options I-II, respectively. According to Figs. 8.14 and 8.15, with the increase of the number of missing PMUs, robust ensemble IS and DTSS show the different degree of decline in accuracy. By contrast, the GAN-based IS is not significantly influenced, and still can maintain a relatively high SA accuracy.

(3) *Average Accuracy against PMU Missing Comparison*: In this part, average SA accuracy index (ADAI) is utilized to quantify the SA accuracy and robustness for incomplete PMU measurement:

$$ADAI = \frac{1}{M}\frac{1}{F_c}\sum_{f=1}^{F_c}\sum_{m=1}^{M}A_m \times 100\% \qquad (8.16)$$

where F_c represents the number of different contingencies. A_m is the average SA accuracy with m PMU missing, and ADAI is the average SA accuracy under all PMU missing conditions.

A larger ADAI value implies higher SA accuracy and robustness against incomplete PMU measurements. Two state-of-the-art methods are also compared in Table 8.7. It can be seen that the developed GAN-based IS always has the highest accuracy and thus robustness against missing-data conditions under the PMU placement options I-II among three methods.

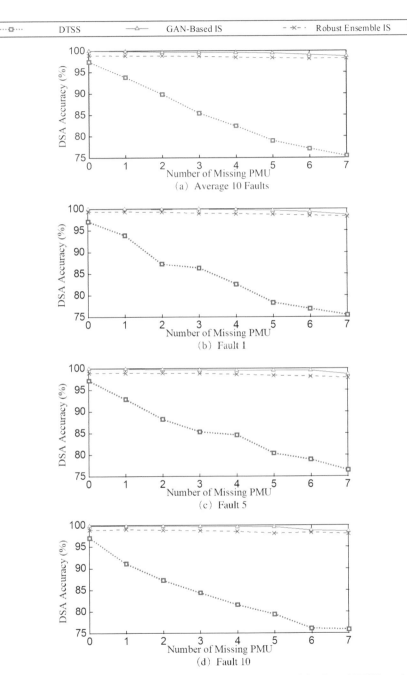

Fig. 8.14: SA testing results of GAN-based IS, robust ensemble IS and DTSS under the PMU placement option I: (a) average on 10 faults, (b) fault 1, (c) fault 5, and (d) fault 10

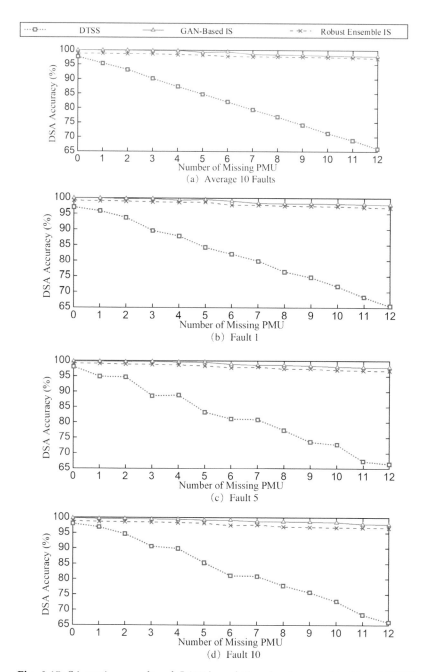

Fig. 8.15: SA testing results of GAN-based IS, robust ensemble IS and DTSS under the PMU placement option II: (a) average on 10 faults, (b) fault 1, (c) fault 5, and (d) fault 10

Table 8.7: ADAI results and computational efficiency of different methods

Method	Computational efficiency (No. of classifiers)		ADAI	
	PMU Option I	PMU Option II	PMU Option I	PMU Option II
GAN-based IS	1	1	99.40%	99.04%
Robust ensemble IS	19	36	98.48%	97.96%
DTSS [67]	1	1	83.28%	80.81%
Feature Eestimation [173]	255	8191	96.99%	96.12%

(4) *Computational Efficiency Comparison* : Another significant advantage is the computational efficiency. The GAN-based IS can fill up the incomplete PMU data under any PMU missing scenario by training only one classifier. But the existing methods need to train more classifiers to be practical, for N PMUs, the total number of classifiers could be $2^N - 1$ at most. The number of classifiers needed to perform the SA is also compared in Table 8.7, where it can be seen that the developed GAN-based IS is greatly computational efficient than the existing methods which require many more classifiers to be trained.

8.7 A Missing-data Tolerant IS of Real-time STVS Prediction

The concept of power grid observability has been explored in Section 8.2 and a robust ensemble IS is developed to sustain the SA performance during different PMU loss (i.e. PMU becomes unavailable) events. In Section 8.4, a PMU clustering algorithm is proposed to compose a comprehensive input clusters that can sustain optimal observability of the IS for different PMU-loss scenarios. Based on such PMU cluster-based inputs, a high SA accuracy can be withheld for a large number of lost PMUs. Moreover, in Section 8.5, a fully data-driven IS, based on GAN, is developed to directly and accurately fill up the missing data without depending on PMU observability and network topologies and which is more generalized and extensible.

However, the previous works are all for pre-fault transient rotor angle SA; the post-disturbance STVS prediction problem is not systematically studied. Besides, a fixed power grid topology is assumed. Yet, in practice, subject to different fault types, it is very likely that the grid topology is changed after the disturbance, e.g. line tripping to clear a short-circuit fault. In this case, the ISs in Sections 8.2-8.6 are not applicable anymore.

To the best of the authors' knowledge, there has not been an effective method reported for post-disturbance STVS prediction with missing-data under fault-induced topology change.

This section aims to fill this gap by developing a missing-data tolerant IS. At the input composition stage, an observability-oriented bus group set is proposed to maintain a high level of grid observability and compose the feature inputs for the SP model under any missing-data scenario. To fully consider the impact of PMU loss and topology change, the proposed bus group set is constructed in a sequential 'add-on' manner, where a basic subset covers the full network topology and multiple add-on subsets cover different post-disturbance topology scenarios.

At the model training stage, a structure-adaptive ensemble model is designed for real-time STVS prediction under missing-data conditions. Depending on the real-time grid observability, the ensemble structure is self-adaptive to use the available feature inputs for real-time STVS prediction. In this section, the structure-adaptive ensemble model is constructed by RVFL units, but other high-performance learning algorithms are also applicable.

Compared with existing methods, the developed IS in this section possesses the following merits:

(1) It is more robust against the missing-data issues that stem from both PMU loss and fault-induced topology change and can sustain a much higher STVS prediction accuracy under a large portion of missing-data inputs.
(2) While most of the existing methods are based on missing data imputation, the developed IS in this section focuses on available data exploitation and is open to any high-performance learning algorithms.
(3) Although this section focuses on STVS prediction, the developed IS can also be used to mitigate the missing-data issues in other post-fault SP problems (e.g. rotor-angle stability, frequency stability).

8.7.1 Practical Post-disturbance Missing-data Scenarios

This section aims to counteract different missing-data issues in post-disturbance STVS prediction. Before developing the IS, the practical missing-data scenarios at the post-disturbance stage need to be understood and analyzed.

For a PMU-monitored power grid, we model the physical power grid and the measurement system using two undirected graphs – $\mathcal{G}_g(\mathbf{E}, \boldsymbol{\Theta})$ and $\mathcal{G}_m(\mathbf{M}, \mathbf{D}, \boldsymbol{\Lambda})$, where $\mathbf{E}, \boldsymbol{\Theta}, \mathbf{M}, \mathbf{D}$ and $\boldsymbol{\Lambda}$ are the sets of buses, branches, PMUs, PDC and communication links, respectively. Based on the observability concept introduced in Section 8.2.1, we denote the observability of a set of PMUs $\mathbf{p} = \{p_1, p_2, ..., p_k\} \subseteq \mathbf{M}$ as $O(\mathbf{p}) = \{\mathbf{i}_\mathbf{p}, \mathbf{n}_\mathbf{p}, \mathbf{z}_\mathbf{p}\} \subseteq \mathbf{E}$, where $\mathbf{i}_\mathbf{p}$ includes the installation buses of the PMUs in $\mathbf{p}$, $\mathbf{n}_\mathbf{p}$ includes neighboring buses of

i_p, and z_p includes the other observable buses owing to ZIB effect. If all the buses in the system are observable, the observability is called *complete* (i.e. $O(M) = E$), otherwise *incomplete* (i.e. $O(M) \subset E$). Achieving complete observability means that the voltage data of all the buses in the system is available for STVS prediction.

Most of the existing STVS prediction methods assume that complete observability is available to the SP model. But in practice, some unintended events can lead to incomplete observability, which requires the SP model to be able to assess the voltage stability under a missing-data condition.

In practice, the missing-data events take place in both measurement systems and physical power grid. Some typical missing-data events in measurement systems are described below and which can be seen as PMU loss events.

(1) *PMU failure*
Failure of a PMU means that the measurements from that PMU become unavailable for STVS prediction, which is equivalent to the loss of that PMU from measurement system.

(2) *PDC failure*
A PDC works for collecting the PMU measurements in an area and transmitting them to the control center. When a PDC fails, the control center will lose its downstream measurements, which can be seen as the loss of the downstream PMUs.

(3) *Communication latency*
According to IEEE Standard C37.118.2-2011 (I.S. Association, 2011), the typical communication latency between PMU and PDC is 20 ms to 50 ms, but it may be even longer due to temporary data congestion. If any measurement cannot be received within an acceptable time (e.g. 50 ms), the STVS prediction should still be carried out without further waiting. The impact of communication latency can be seen as the loss of the sourced PMUs of those delayed measurements.

Mathematically, suppose a set of PMUs p_m are missing in a PMU loss event e_m, then the measurement system becomes $G_m \mid e_m = G_m(M', D, \Lambda)$, where $M' = M \setminus p_m$, and the grid observability is $O(M) \mid e_m = O(M') = \{i_{M'}, n_{M'}, z_{M'}\} \subseteq O(M)$.

In the physical power grid, the typical missing-data event is the topology change, e.g. transmission line tripping actions to clear the fault. Suppose an outage of a set of branches l_m in a topology change event e_g, then $G_g \mid e_g = G_g(E, \Theta')$, where $\Theta' = \Theta \setminus l_m$. A topology change in Θ can change the neighboring relationship between buses, which affects the grid observability, i.e. $O(M) \mid e_g = \{i_M, n_M \mid \Theta', z_M \mid \Theta'\} \subseteq O(M)$. Real-time STVS prediction is a post-disturbance activity that is exposed to both PMU loss and fault-induced topology change events. When the power grid is experiencing e_m and e_g simultaneously, the grid observability becomes $O(\mathbf{M}) \mid e_m, e_g = O(\mathbf{M}') \mid e_g = \{\mathbf{i}_{M'}, \mathbf{n}_{M'} \mid \Theta', \mathbf{z}_{M'} \mid \Theta'\} \subseteq O(\mathbf{M})$.

An example on a nine-bus system is shown in Fig. 8.16 to demonstrate the impact of different missing-data scenarios on power grid observability. In Fig. 8.16(a), complete observability is achieved by two PMUs, $\mathbf{M} = \{p_1, p_2\}$, where each of which can observe a small section of the grid: $O(p_1) = \{\mathbf{i}_{p_1}, \mathbf{n}_{p_1}\} = \{b_4, b_5, b_6\}$ and $O(p_2) = \{\mathbf{i}_{p_2}, \mathbf{n}_{p_2}\} = \{b_2, b_7, b_8, b_9\}$, where b_i denotes the bus i. Owing to the ZIB effect, the combination of two PMUs can observe an even larger area that covers the whole grid, $O(\mathbf{M}) = \{\mathbf{i}_M, \mathbf{n}_M, \mathbf{z}_M\} = E$, where $\mathbf{i}_M = \mathbf{i}_{p_1} \cup \mathbf{i}_{p_2}$, $\mathbf{n}_M = \mathbf{n}_{p_1} \cup \mathbf{n}_{p_2}$, and $\mathbf{z}_M = \{b_1, b_3\}$.

In Fig. 8.16(b), the remaining grid observability under three missing-data events is shown. The first event e_1 is a PMU loss event $\mathbf{p}_m = \{p_1\}$,

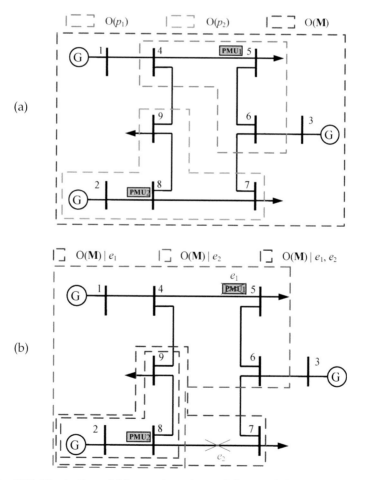

Fig. 8.16: Illustration of (a) complete observability, and (b) incomplete observability under different missing-data scenarios

so $\mathbf{M}' = \mathbf{M}\backslash p_1 = \{p_2\}$, meaning the remaining grid observability equals to the observability of p_2, i.e. $O(\mathbf{M}) \mid e_1 = O(p_2)$. The second event e_2 is a topology change event $\mathbf{l}_m = \{l_{7,8}\}$, so $\Theta' = \Theta \backslash l_{7,8}$. With $l_{7,8}$ tripped, b_7 is not a neighboring bus of $\mathbf{i}_{p2} = b_8$, so $\mathbf{n_M} \mid \Theta' = \mathbf{n_M} \backslash b_7$. Without the observability on b_7, we cannot further calculate the voltage phasor of $b_{3;}$ so $\mathbf{z_M} \mid \Theta' = \mathbf{z_M} \backslash b_3$. The third event is the co-existence of e_1 and e_2, which incorporates the impact of PMU loss and topology change. In this case, the grid observability becomes even smaller than the original observability of p_2, $O(\mathbf{M}) \mid e_1, e_2 = \{\mathbf{i}_{p_2}, \mathbf{n}_{p_2} \mid \Theta'\} = \{b_2, b_8, b_9\} \subset O(p_2)$.

8.7.2 Impact of Missing-data Events

The conventional IS-based STVS prediction models require complete feature inputs, and missing-data events will result in model invalidity if any feature input becomes unavailable. For existing methods devised to mitigate missing-data issues in (T.Y. Guo and J.V. Milanovic, 2013; M. He, *et al.*, 2013) and previous sections, they remain valid in the case of missing data, but inevitably experience accuracy impairment due to lack of feature inputs. Generally, a larger amount of missing data leads to more significant impairment in accuracy. The aim of this section is to improve the missing-data tolerance of real-time STVS prediction, i.e. sustain a higher level of accuracy under the different missing-data scenarios from PMU loss and fault-induced topology change.

8.7.3 A Review on PMU Clustering Algorithm

In Section 8.4.2, we have proposed an analytical algorithm to compose different PMU clusters to ensure feasible feature inputs in the SA model. As mathematically proved, this method can minimize the number of PMU clusters to maximize the grid observability under any PMU loss event. However, this algorithm is designed for a fixed network topology and not suitable for post-disturbance SP tasks. A review and discussion on this PMU clustering algorithm is provided in the sequel.

For a power grid with N PMUs, the number of possible PMU combinations is $2^N - 1$. The principle of the PMU clustering method is to select the minimum number of PMU combinations to ensure that, for any PMU loss scenario, the combined observability of the PMU clusters that remain complete can fully cover the remaining grid observability. Mathematically, the PMU clusters serve as the solution for the following optimization problem:

$$\underset{\mathcal{P}}{\text{Min}}|\mathcal{P}| \qquad (8.17)$$

s.t. $$O(\mathbf{p}) = \bigcup\nolimits_{\mathbf{c}_j \in \mathcal{V}} O(\mathbf{c}_j), \forall \mathbf{p} \subset \mathbf{M} \qquad (8.18)$$

where

$$\mathcal{V} = \{\mathbf{c} \in \mathcal{P} \mid \mathbf{c} \subseteq \mathbf{p}\} \tag{8.19}$$

where $\mathbf{p}$ represents the available PMUs, $\mathcal{P}$ is the collection of PMU clusters, and $\mathcal{V}$ is the collection of the complete PMU clusters during the PMU loss event.

The detailed PMU clustering process for a N_b-bus power grid is mathematically presented as follows. For each bus b_i, all the non-redundant PMU combinations that can observe b_i are searched and collected in a non-redundant PMU combination set $\mathcal{T}_{b_i}$

$$\mathcal{T}_{b_i} = \{\mathbf{c}_j \subseteq \mathbf{M} \mid b_i \in O(\mathbf{c}_j), b_i \notin O(\mathbf{c}_k), \forall \mathbf{c}_k \subset \mathbf{c}_j\} \tag{8.20}$$

Then the final PMU clustering result $\mathcal{P}$ is the union of $\mathcal{T}_{b_i}$ over all the buses in the system

$$\mathcal{P} = \bigcup\nolimits_{b_i \in \mathbf{E}} \mathcal{T}_{b_i} \tag{8.21}$$

According to (8.20), for a PMU combination $\mathbf{c}_j$ that can observe bus b_i, $\mathbf{c}_j$ is called non-redundant if none of the subsets of $\mathbf{c}_j$ can observe b_i. In this definition, $\mathbf{c}_j$ tends to be the PMU combination with the lowest cardinality to observe b_i, which can cover most PMU loss scenarios (i.e. remain complete for most PMU loss scenarios). By only collecting non-redundant PMU combinations, a minimum number of clusters is required to cover all the PMU loss scenarios.

In data-driven SA, the SA model consists of multiple machine learning units, and the data measurement set from each PMU cluster is used as the feature inputs for a machine learning unit. Since the number of PMU clusters is minimized according to (8.17), the number of machine learning units is also minimized. This can significantly reduce the overall computation complexity. In the case of a PMU loss event, only the machine learning units supplied with complete feature inputs are available for SA. In this situation, the PMU clusters ensure that the aggregated feature inputs to the available machine learning units fully cover the remaining grid observability according to (8.18), which maximizes the exploitation of available data. Compared to other methods based on the learning algorithms (T.Y. Guo and J.V. Milanovic, 2013; M. He, *et al.*, 2013), the PMU clustering algorithm focuses on model input composition, which can better sustain the SA accuracy during PMU loss events.

Although the PMU clustering algorithm shows its strength in mitigating the PMU loss events, it is designed only for pre-fault SA where a full network topology is assumed. However, subject to a topology change, the observability of the PMUs can be affected (e.g. in Fig. 8.16(b), the topology change e_2 shrinks the observability of p_2), so a PMU clustering result is not robust to different network topologies.

8.7.4 Observability-oriented Bus Group Set

To fill the gap, an observability-oriented bus group set is proposed in this section to maintain the grid observability under different PMU loss and/or topology change scenarios. The voltage measurements on the bus group set are used as the feature inputs for the SP model.

(1) *Working principle*

As illustrated in Fig. 8.17, the observability-oriented bus group set $\mathcal{G}_b$ rearranges the buses in the system into multiple bus groups and it is constructed under a sequential 'add-on' structure. Mathematically, $\mathcal{G}_b$ is the union of a basic subset $\mathcal{B}$ and multiple add-on subsets $\{\mathcal{A}_t, t = 1, 2,...., N_T\}$ as follows:

$$\mathcal{G}_b = \mathcal{B} \cup \mathcal{A}_1 \cup \mathcal{A}_2 \cup \cdots \cup \mathcal{A}_{N_T} \tag{8.22}$$

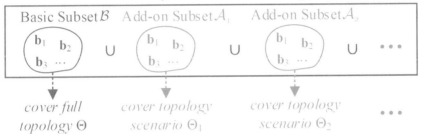

Fig. 8.17: Sequential 'add-on' structure of observability-oriented bus group set

Each subset is a collection of multiple bus groups. The basic subset $\mathcal{B}$ only mitigates the PMU loss events under full network topology $\boldsymbol{\Theta}$, and each add-on subset $\mathcal{A}_t$ serves as a supplementary to $\mathcal{B}$ to mitigate the PMU loss events under a topology scenario $\boldsymbol{\Theta}_t$. Such topology scenarios represent the possible topology changes induced by faults (i.e. transmission line tripping to clear the fault). By combining the basic subset and the add-on subsets, the bus groups can handle different PMU loss events under various topology scenarios.

The advantage of such a sequential 'add-on' structure is the high flexibility to satisfy different power system operation needs. For instance, if one more topology scenario $\boldsymbol{\Theta}_{N_T+1}$ needs to be considered, a new add-on subset $\mathcal{A}_{N_T+1}$ can be quickly calculated and integrated to the original bus group set to cover the new topology scenario, i.e. $\mathcal{G}_b = \mathcal{G}_b \cup \mathcal{A}_{N_T+1}$. In this sense, the observability-oriented bus group set is able to mitigate the missing-data issue under any topology change.

(2) *Development of basic subset*

The basic subset consists of bus groups that are the observability of the PMU clusters derived using the Algorithm 8.1 in Section 8.4.2.

$$B = \{\mathbf{b}_i \mid \mathbf{b}_i = O(\mathbf{c}_i), \forall \mathbf{c}_i \in \mathcal{P}\} \tag{8.23}$$

As mentioned before, the PMU clustering algorithm does not consider the impact of topology changes, so B can only cover the PMU loss scenarios under full topology. The bus groups in B are called *basic bus groups*. Since the basic bus groups are directly developed from PMU clusters, the optimality of PMU clusters (i.e. (8.17)-(8.19)) also applies to basic bus groups, meaning that the basic subset is composed of a minimum number of bus groups to maximize the grid observability under full topology.

(3) *Development of an add-on subset*

A straightforward strategy of developing an add-on subset is to repeat the same procedure of developing basic subset under the topology change scenario. However, such a strategy requires an exhaustive search over all the buses in the system. This suffers from high computation complexity, especially for large-scale power grids with numerous topology scenarios. To improve the computation efficiency, we propose to first identify the critical buses that are subject to observability change under the new topology, and then search new bus groups only over the critical buses. In doing so, the scale of the searching problem can be dramatically diminished, which reduces the computation complexity of developing an add-on subset. In summary, an add-on subset is obtained via two steps as shown in Fig. 8.18 – critical buses identification and new bus group searching.

Critical buses identification - In the basic subset, the buses are grouped together since they are observable by a PMU cluster under full topology. However, following a topology change, the observability of some PMUs may change. This means that the buses originally in a group may not still be observable by the same PMU cluster. As such, we identify the buses subject to observability change as critical buses and compose new bus groups to accommodate them. In a topology change event $\mathbf{\Theta}_t = \mathbf{\Theta} \setminus \mathbf{l}_m$, the PMU combinations that can observe the terminal buses of the tripped lines may change. So the bus groups involving those terminal buses need to be re-analyzed. In this sense, the critical buses are technically defined as: *under a topology change, the critical buses are the buses in basic bus groups that involve terminal buses of the tripped transmission lines*

$$\mathbf{b}_t^* = \{b_k \in \mathbf{b}_c \in B \mid \mathbf{b}_c \cap \mathbf{n}_{l_m} \neq \varnothing\} \tag{8.24}$$

where

$$\mathbf{n}_{l_m} = \{b_i, b_j \mid l_{i,j} \in \mathbf{l}_m\} \tag{8.25}$$

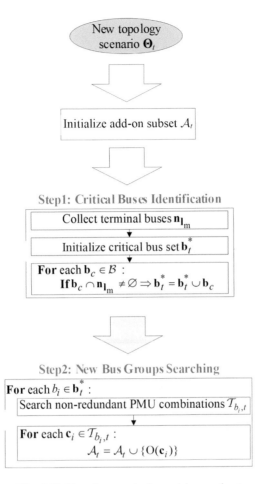

Fig. 8.18: Development of an add-on subset

where $\mathbf{b}_t^*$ is the set of critical buses and $\mathbf{n}_{1_m}$ represents the set of terminal buses of $\mathbf{l}_m$.

New bus group searching – Based on the critical buses $\mathbf{b}_t^*$ and the observability under the new topology scenario $\mathbf{\Theta}_t$, we search the new bus groups via a similar process as for the basic subset. First, for each bus $b_i \in \mathbf{b}_t^*$, the non-redundant PMU combinations that can observe b_i under the topology scenario $\mathbf{\Theta}_t$ are searched and collected in $\mathcal{T}_{b_i,t}$

$$\mathcal{T}_{b_i,t} = \{\mathbf{c}_j \subseteq \mathbf{M} \mid b_i \in \mathrm{O}(\mathbf{c}_j \mid \mathbf{\Theta}_t), b_i \notin \mathrm{O}(\mathbf{c}_k \mid \mathbf{\Theta}_t), \forall \mathbf{c}_k \subset \mathbf{c}_j\} \qquad (8.26)$$

Then, new bus groups are composed as the observability of the PMU combinations in $\mathcal{T}_{b_i,t}$, and the add-on subset $\mathcal{A}_t$ collects the new bus groups over all critical buses:

$$\mathcal{A}_t = \{\mathbf{b}_i \mid \mathbf{b}_i = O(\mathbf{c}_i \mid \boldsymbol{\Theta}_t), \forall \mathbf{c}_i \in \mathcal{P}_t^*\} \tag{8.27}$$

where

$$\mathcal{P}_t^* = \bigcup_{b_i \in \mathbf{b}_t^*} \mathcal{T}_{b_i, t} \tag{8.28}$$

Since the new bus groups are developed from non-redundant PMU combinations (similar to the development of basic subset), the add-on subsets would minimize the number of new bus groups to maximize grid observability under $\boldsymbol{\Theta}_t$.

8.7.5 Preliminary Tests on Benchmark Power Systems

The proposed observability-oriented bus group set is tested on several benchmark power systems – IEEE 14-bus, IEEE 24-bus, IEEE 30-bus, and New England 39-bus systems. The PMU locations are listed in Table 8.8 (S. Chakrabarti and E. Kyriakides, 2008). Under ZIB effect, such PMU placement achieves complete observability with a minimum number of PMUs. The selected topology scenarios for each benchmark system include all the scenarios of tripping single ($N - 1$) and double ($N - 2$) transmission lines without any islanding effect. The test is performed on a PC with 3.4GHz CPU and 16GB RAM.

Table 8.8: PMU locations on benchmark power systems

Power systems	Located buses
IEEE 14-bus system	b_2, b_6, b_9
IEEE 24-bus system	$b_2, b_8, b_{10}, b_{15}, b_{20}, b_{21}$
IEEE 30-bus system	$b_1, b_2, b_{10}, b_{12}, b_{15}, b_{19}, b_{27}$
New England 39-bus system	$b_3, b_8, b_{10}, b_{16}, b_{20}, b_{23}, b_{25}, b_{29}$

The case of IEEE 14-bus system is elaborated in detail to demonstrate how the observability-oriented bus group set is derived. The power grid and the installed PMUs are shown in Fig. 8.19. With three PMUs, there are seven possible PMU combinations: $\mathcal{C} = \{\{p_1\}, \{p_2\}, \{p_3\}, \{p_1, p_2\}, \{p_1, p_3\}, \{p_2, p_3\}, \{p_1, p_2, p_3\}\}$. To develop the basic subset, we need to first identify the non-redundant PMU combinations $\mathcal{T}_{b_i}$ from $\mathcal{C}$; for example, the candidate PMU combinations that can observe b_1 are $\{p_1\}, \{p_1, p_2\}, \{p_1, p_3\}$, and $\{p_1, p_2, p_3\}$. Among them, only $\{p_1\}$ is non-redundant by definition because the other PMU combinations include $\{p_1\}$ as a subset (i.e. (8.20) is violated). We suppose p_2 and p_3 are lost simultaneously in a PMU loss event. Then $\{p_1\}$ remains available to provide b_1 observability to its corresponding machine learning unit. However, the other candidate PMU combinations become incomplete and thereby are unable to provide observability. This

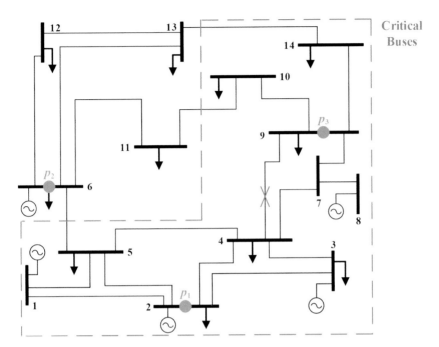

Fig. 8.19: IEEE 14-bus system with three PMUs

example verifies that $\{p_1\}$, as a non-redundant PMU combination for b_1, can cover the most PMU loss scenarios. After scanning all the buses, all the non-redundant PMU combinations are collected as PMU clusters $\mathcal{P}$ = $\{\{p_1\}, \{p_2\}, \{p_3\}\}$. Then, the basic subset is composed as $\mathcal{B} = \{O(p_1), O(p_2), O(p_3)\} = \{\{b_1, b_2, b_3, b_4, b_5\}, \{b_5, b_6, b_{11}, b_{12}, b_{13}\}, \{b_4, b_7, b_8, b_9, b_{10}, b_{14}\}\}$.

The next stage is to develop the add-on subset for each topology scenario. Here, we take the tripping of $l_{4,9}$ (i.e. $\mathbf{\Theta}_t = \mathbf{\Theta} \setminus l_{4,9}$) as an example to show how the new bus groups are composed for a topology change. With $l_{4,9}$ tripped, the terminal buses are b_4 and b_9 which are included in the basic bus groups $O(p_1)$ and $O(p_3)$. So the buses in $O(p_1)$ and $O(p_3)$ are identified as critical buses (as shown in Fig. 8.19). Then, the non-redundant PMU combinations $\mathcal{T}_{b_j,t}$ are searched over each critical bus in $\mathbf{b}_t^*$. In this process, we successfully prevent the searching over the buses in $O(p_2)$, which reduces the computation complexity. As a result, the collected $\mathcal{T}_{b_j}$ is $\mathcal{P}_t^* = \{\{p_1\}, \{p_3\}, \{p_1, p_3\}\}$, where $\{p_1, p_3\}$ is established to accommodate b_8 because b_8 is only observable when p_1 and p_3 are simultaneously available under the new topology. The add-on subset $\mathcal{A}_t$ is composed of the observability of each $\mathcal{T}_{b_j,t}$ under topology scenario $\mathbf{\Theta}_t$: $\mathcal{A}_t = \{O(p_1 \mid \mathbf{\Theta}_t), O(p_3 \mid \mathbf{\Theta}_t), O(p_1, p_3 \mid \mathbf{\Theta}_t)\} = \{\{b_1, b_2, b_3, b_4, b_5\}, \{b_7, b_9, b_{10}, b_{14}\}, \{b_1, b_2, b_3, b_4, b_5, b_7, b_8, b_9, b_{10}, b_{14}\}\}$. The basic subset $\mathcal{B}$ together with add-on subset $\mathcal{A}_t$ covers the PMU loss scenarios under topology scenario $\mathbf{\Theta}_t$. By repeating above procedures, the add-on subset for other topology scenarios can also be developed.

For each benchmark system, the statistics of observability-oriented bus group set is listed in Table 8.9. It can be seen that, as the size of power grid increases, more topology scenarios have to be considered, which increases the number of developed bus groups and the computation complexity. However, the observability-oriented bus group set is established off-line and serves as a generic permanent solution to mitigate different missing-data scenarios. Thus the computation complexity of developing the bus groups does not impact the real-time STVS prediction efficiency. Moreover, the sequential 'add-on' process of bus groups is illustrated in Fig. 8.20. For more topology scenarios, the number of bus groups increases, but the increasing trend flats out at some topology scenarios.

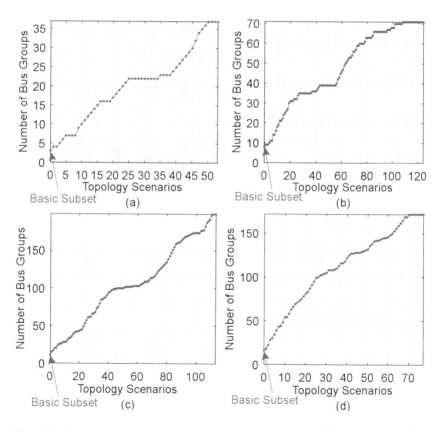

Fig. 8.20: Sequential 'add-on' process of bus groups for (a) IEEE 14-bus, (b) IEEE 24-bus, (c) IEEE 30-bus, and (d) New England 39-bus systems

Table 8.9: Observability-oriented bus group set for different systems

Power systems	No. of topology scenarios	No. of basic bus groups	Total no. of bus groups	CPU time (s)
IEEE 14-bus	53	3	37	2.35
IEEE 24-bus	124	8	71	62.80
IEEE 30-bus	113	11	199	369.22
New England	77	19	172	1109.77

8.7.6 Structure-adaptive Ensemble Model

Conventionally, the ensemble learning model aggregates the machine learning units under a fixed structure and assumes all feature inputs to be available. However, under a missing-data condition, some feature inputs will become unavailable. Then the ensemble learning model may be unable to work. To avoid such an impact, a structure-adaptive ensemble learning model is developed. It can adapt its structure to use the available feature inputs under different missing-data scenarios.

The structure-adaptive ensemble model is illustrated in Fig. 8.21. The feature inputs to each RVFL consists of the voltage measurement on each bus group, and the STVS prediction decision is a weighted aggregation of the RVFL outputs. Suppose there are K bus groups in the observability-oriented bus group set, then K RVFLs are needed to construct the ensemble model. On top of the conventional ensemble learning structure, the unique feature of the developed model is its real-time structure adaption to the available feature inputs. Due to the change of PMU status and network topology, the grid observability can change in real-time. Based on such real-time grid observability, the structure of the ensemble model is controlled by an ensemble structure signal L which is a K-bit binary number. Each bit of L controls the availability of each RVFL. With

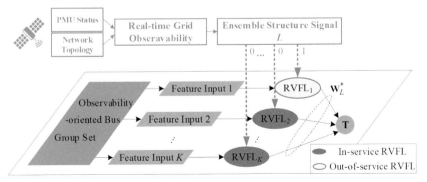

Fig. 8.21: Structure-adaptive ensemble learning model

complete observability (i.e. full data availability), all the bits of L are set to 0, meaning all the RVFLs are in-service. With incomplete observability (i.e. missing-data condition), the corresponding bits for the RVFLs with incomplete feature inputs are set to 1; meaning that those RVFLs are out-of-service. Under such control, the unavailable feature inputs are avoided in ensemble classification, which achieves real-time structure adaption. Moreover, since the observability-oriented bus group set maintains the grid observability under any PMU loss and/or topology change scenario according to (8.18), the in-service RVFLs will fully exploit the available feature inputs to make ensemble decision, which sustains STVS prediction accuracy.

The off-line training and the on-line structure adaption of the developed ensemble model are elaborated below:

- **Off-line Training**

The training of the structure-adaptive ensemble model is divided into two stages – single RVFL training and ensemble training. The principle is to train every single RVFL first, and then train the whole ensemble based on the training output from every single RVFL.

(1) *Single RVFL training*
Given a database $\mathcal{D}$ of N training samples $\{(\mathbf{x}_1, t_1), \ldots, (\mathbf{x}_N, t_N)\}$ and the observability-oriented bus group set $\mathcal{G}_\mathrm{b} = \{\mathbf{b}_1, \ldots, \mathbf{b}_K\}$, a single RVFL is trained following Algorithm 8.4.

Algorithm 8.4: Single RVFL Training
For $j = 1$ to K:
    Step 1: Extract the feature space from $\mathcal{D}$ according to $\mathbf{b}_j$ and construct a new dataset $\mathcal{D}_j$.
    Step 2: Randomly select the number of hidden nodes from the optimal range $[h_{min}, h_{max}]$.
    Step 3: Train a RVFL r_j as a regression model using $\mathcal{D}_j$ and obtain the training output $r_j(\mathbf{x}_i)$, $i = 1, 2, \ldots, N$.
End For

---

(2) *Ensemble training*
The STVS prediction is modeled as a binary classification problem, i.e. stable or unstable. In the developed ensemble model, every single RVFL is first trained as a regression model (i.e. the output is a continuous value). Then, the STVS status is mapped as the weighted sum of individual RVFL outputs

$$t_i = \sum_{j=1}^{K} \omega_j r_j(\mathbf{x}_i), \ i = 1, 2, \ldots, N \tag{8.29}$$

where ω_j is the ensemble weight on jth RVFL. Here, all the RVFL outputs $r_j(\mathbf{x}_i)$ can be concatenated as a training output matrix $\mathbf{R}$, the ensemble weights can be concatenated as an ensemble weight vector **W**, and the class label of training instances can be concatenated as a class label vector **T**

$$\mathbf{R} = \begin{bmatrix} r_1(\mathbf{x}_1) & r_2(\mathbf{x}_1) & \cdots & r_K(\mathbf{x}_1) \\ r_1(\mathbf{x}_2) & r_2(\mathbf{x}_2) & \cdots & r_K(\mathbf{x}_2) \\ \vdots & \vdots & \ddots & \vdots \\ r_1(\mathbf{x}_N) & r_2(\mathbf{x}_N) & \cdots & r_K(\mathbf{x}_N) \end{bmatrix} \tag{8.30}$$

$$\mathbf{W} = \begin{bmatrix} \omega_1 & \omega_2 & \cdots & \omega_K \end{bmatrix}^T \tag{8.31}$$

$$\mathbf{T} = \begin{bmatrix} t_1 & t_2 & \cdots & t_N \end{bmatrix}^T \tag{8.32}$$

Then (8.29) can be rewritten in a matrix form:

$$\mathbf{RW} = \mathbf{T} \tag{8.33}$$

The ensemble training is to find a least-squares solution $\mathbf{W}^*$ for the linear system in (8.33). Similar to single RVFL training, $\mathbf{W}^*$ is also computed using Moor-Penrose pseudo-inverse to pursue the smallest norm:

$$\mathbf{W}^* = \mathbf{R}^\dagger \mathbf{T} \tag{8.34}$$

- **On-line Structure Adaption**

  Depending on the real-time grid observability, the availability of each RVFL is controlled by the ensemble structure signal L. With complete observability, all the RVFLs in the ensemble model are in-service. Thus the STVS prediction decision can be simply made as:

$$\hat{t}_- = \mathbf{R}_- \times \mathbf{W}^* \tag{8.35}$$

where

$$\mathbf{R}_- = \begin{bmatrix} r_1(\mathbf{x}_-) & \cdots & r_K(\mathbf{x}_-) \end{bmatrix} \tag{8.36}$$

where $\mathbf{x}_-$ is a post-fault observation and $\hat{t}_-$ is the STVS prediction decision.

During a missing-data event e, some RVFLs will become out-of-service based on L, which reduces the rank of $\mathbf{R}_-$

$$\text{rank}(\mathbf{R}_- \mid e) < \text{rank}(\mathbf{R}_-) \tag{8.37}$$

In this situation, there will be a rank mismatch between $\mathbf{R}_-$ and the original ensemble weight vector $\mathbf{W}^*$. Thus a new ensemble weight vector must be calculated in real-time to adapt to the rank change of $\mathbf{R}_-$. With missing data, suppose only K_L RVFLs are in-service, then we modify the training output matrix **R** as $\mathbf{R}_L$ by only preserving the columns corresponding to the in-service RVFLs. Following (8.34), the new ensemble weight vector $\mathbf{W}^*_L$ is calculated as:

$$\mathbf{W}^*_L = \mathbf{R}_L{}^\dagger\mathbf{T} \tag{8.38}$$

Since (8.38) is a single-step matrix computation, the burden of re-calculating $\mathbf{W}^*_L$ should satisfy the real-time application need. The STVS prediction decision under missing-data event e is made as:

$$\hat{t}_- \mid e = \mathbf{R}_- \mid e \times \mathbf{W}^*_L \tag{8.39}$$

8.8 Numerical Test for Missing-data Tolerant IS

The missing-data tolerant IS is tested on New England 39-bus system as shown in Fig. 8.22. Considering the impact of RES, three wind farms are integrated to replace the synchronous generators on bus 32, 35, and 37. Also, to accommodate the wind power variation, the capacities of other generators are set to be 1.5 times larger. The wind power penetration in the system is 36.58 per cent. The PMUs are placed according to the PMU Placement Option I in Table 8.2.

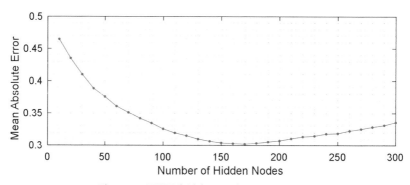

Fig. 8.22: RVFL hidden node tuning result

8.8.1 Database Generation

A comprehensive database is generated considering various pre-fault operating conditions, load models and faults.

(1) *Pre-fault operating condition*
6,536 OPs are generated through OPF calculation under random variation of wind power generation and load demand. The wind power output varies between 0 and its capacity, and load demand varies between 80 per cent and 120 per cent of the base case.

(2) *Induction motor load percentage*
The induction motor load dynamics has a substantial impact on STVS. The percentage of induction motor loads in the system is randomly sampled between 0 per cent and 80 per cent for each OP.

(3) *Fault type and duration time*
For each OP, a three-phase fault is applied on a randomly selected bus with a fault duration randomly sampled between 0.1 s and 0.3 s. The fault is cleared by tripping either a single ($N-1$) or double ($N-2$) transmission lines, which simulates the different topology change scenarios.

Then, the post-disturbance voltage trajectories for each OP are obtained via TDS using TSAT. The simulation step size is 0.1 s, and the total simulation time is 5 s after the fault clearance. If all the bus voltages successfully recover to an acceptable equilibrium (e.g. between 0.9 p.u. and 1.1 p.u.), the instance is labeled as stable; otherwise unstable. In the generated database, 4,023 out of the 6,536 instances are stable and the remaining 2,513 instances are unstable. To test the developed IS, 20 per cent of the 6,536 instances are randomly selected as the testing dataset and the remaining serve as training dataset.

8.8.2 Observation Window

The length of observation window for post-disturbance STVS prediction is an important concern: with longer observation window following a fault, the STVS prediction tends to be more accurate since more voltage trajectory data is available. However, less time will be left for emergency control. In Chapter 6, a self-adaptive decision-making structure is used for post-disturbance SP, which progressively increases the observation window to make an accurate SP as early as possible.

Since this section focuses on the missing-data tolerance, only fixed observation windows are considered. In the test, three different observation windows are selected for STVS prediction, respectively 0.8 s, 1 s, and 1.2 s after the fault clearance. Nevertheless, the developed IS is also immediately available for self-adaptive implementation.

8.8.3 Model Training

Every single RVFL is trained following Algorithm 8.4, and the whole ensemble model is trained using (8.30)-(8.34). In single RVFL training, the feature inputs for each RVFL are decided by the observability-oriented bus group set. The activation function is *sigmoid*, and the number of hidden nodes is tuned as shown in Fig. 8.22 where [150, 180] is selected as the optimal hidden node range because the lowest mean absolute error is achieved within this range. As a result, 172 single RVFLs (i.e. equal to the number of bus groups) are trained for the structure-adaptive ensemble model.

8.8.4 Comparative Study

The developed IS is compared with three existing methods for missing-data problems.

(1) *Mean imputation (MI)* (E. Batista and M.C. Monard, 2003)
MI method is to substitute each missing feature by the mean value of that feature in the training data, which is a baseline method to deal with the missing data problem. The applied SP model is a single RVFL.

(2) *DTSS* (T.Y. Guo and J.V. Milanovic, 2013)
A DTSS is a DT that can substitute the splitting rule on each missing input by its surrogate splitting rule. Such surrogate split is greatly associated with the original split. This method has been applied to mitigate the missing-data problem in power system pre-fault SA.

(3) *Random forest with surrogate split (RFSS)* (L. Breiman, 2001)
A RFSS is an ensemble model of DTSS to improve the accuracy. The number of single DTs in RFSS is set to 172 (i.e. identical to the number of RVFLs in the developed IS).

This comparative study has two parts. The first part is to test the four methods with full data availability. The testing results at different observation windows are listed in Table 8.10 where all methods achieve high STVS prediction accuracy (over 96 per cent). Moreover, all methods show higher accuracy at longer observation window, which indicates the trade-off between observation window and STVS prediction accuracy.

Table 8.10: SP accuracy with full data availability

Methods	Observation windows		
	0.8 s	1 s	1.2 s
MI	96.6%	97.4%	97.7%
DTSS	96.9%	97.1%	97.4%
RFSS	98.1%	98.2%	98.3%
Developed IS	97.8%	98.0%	98.2%

The second part is to compare the performance of the four methods under missing-data conditions. To illustrate the severity of missing-data problem, we gradually increase the percentage of missing data for each observation window. At each missing-data percentage, PMU loss and topology change are incorporated to simulate post-disturbance missing-data condition. The testing results are shown in Fig. 8.23, where the

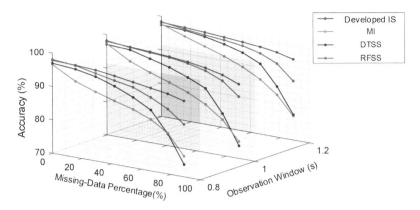

Fig. 8.23: STVS prediction accuracy under different missing-data scenarios

missing-data tolerance of the method is indicated by the slope of accuracy drop which is calculated as below:

$$\bar{s} = \frac{1}{D} \sum_{i=1}^{D} \frac{A_0 - A_i}{d_i} \tag{8.40}$$

where $\bar{s}$ is the average accuracy drop slope, D is the number of missing-data percentage levels (i.e. $D = 7$ in Fig. 8.23), d_i is the ith missing-data percentage value, A_0 is the average accuracy over different observation windows with full data availability, and A_i is the average accuracy at d_i. The calculated $\bar{s}$ of each method is listed in Table 8.11 where a lower $\bar{s}$ indicates stronger missing-data tolerance. As the missing data increases, the other methods for comparison are based on missing input imputation and/or learning algorithm adaption, which sees sharp accuracy drops (i.e. higher $\bar{s}$). In comparison, the developed IS focuses on model input composition and available data exploitation, which sustains the STVS accuracy at the highest level (i.e. lowest $\bar{s}$). Moreover, the developed IS can achieve high classification accuracy under a large portion of missing-data input. For instance, in Fig. 8.23, even with 89.22 per cent missing data, the average accuracy of the developed IS is still as high as 92.74 per cent, 92.22 per cent, and 94.13 per cent at the 0.8 s, 1 s, and 1.2 s observation windows, respectively. Such higher accuracy of the developed IS under missing-data conditions demonstrates its excellent missing-data tolerance in post-disturbance STVS prediction.

Table 8.11: Average accuracy drop slope

Methods	MI	DTSS	RFSS	Developed IS
Slope $\bar{s}$	0.1943	0.1135	0.0812	0.0353

8.9 Conclusion

In practice, the data collected by PMUs in the system can be incomplete. Based on the incomplete PMU data, the performance of power system stability analysis will be impaired or even deteriorated. This chapter focuses the missing-data issue in power system SA, and develops four data-driven IS as mitigations. Those ISs are able to sustain the on-line and real-time SA performances when the received PMU measurement data is incomplete. This improves the SA robustness against the missing-data issues under practical measurement conditions.

First, considering the missing-data issue, this chapter develops a robust ensemble IS for on-line SA, which is adaptable to all possible PMU missing scenarios with negligible degradation on SA accuracy. The robust ensemble IS is built, based on observability-constrained feature subsets and RVFL ensemble model. Each RVFL in the ensemble is trained by the features observable in presence of a single or a combination of PMUs in order to ensure the complete robustness of the whole ensemble model.

Second, this chapter upgrades the robust ensemble IS by replacing the feature sampling process with a strategically designed PMU clustering process. The upgraded IS is more mathematically rigorous. As mathematically proved, under any PMU missing scenario, the upgraded IS is able to ensure minimum computation complexity with maximum grid observability from available PMUs. In doing so, excellent SA robustness against incomplete PMU measurement is achieved.

Third, a GAN-based IS is developed which aims to maintain the SA accuracy and the system stability under any PMU missing conditions without considering PMU observability. Being fully data-driven, the IS does not depend on the PMU observability and/or network topologies. During on-line application, system operators only need to input the practical PMU measurements without considering the detail missing measurements information to select the model or generated data. Therefore, this GAN-based IS offers the system operators more flexibility in manipulating the assessment performance, making the IS more practical in real-world assessment against the PMU missing conditions. By marked contrast, the state-of-the-art methods have much higher computational complexity. The GAN-based IS can only utilize one classifier to maintain a high SA accuracy.

At last, in real-time STVS prediction, the post-fault voltage measurement data can be missing in case of PMU loss events and/or fault-induced topology changes, which can deteriorate the performance of conventional intelligent models. To address this issue, a missing-data tolerant IS is developed for real-time STVS prediction. At the input composition stage, the observability-oriented bus group set can maintain the grid observability for the IS under any missing-data. At the model training stage, the structure-adaptive ensemble model, once constructed, can adapt its structure to use the available feature inputs. Compared with the existing methods, this IS is more tolerant to missing-data issues (covering both PMU loss and fault-induced topology change) and can sustain a much higher STVS prediction accuracy under a large portion of missing data.

All the ISs developed in this chapter have been tested on New England 39-bus system and the testing results have shown promising on-line and real-time SA robustness against the missing-data events.

References

1. Y. Enerdata (2017). Global Energy Statistical Yearbook 2017, Enerdata.
2. C.E. Regulator (2012). About the renewable energy target, Australian Government, 2012.
3. J. Trudeau, B. Obama and E.P. Nieto (2016). North American climate, clean energy, and environment partnership action plan. The White House. Vol. 29.
4. E. Commission (2012). Energy Roadmap 2050, Publications Office of the European Union.
5. M. Wright, P. Hearps and M. Wright (2010). Zero Carbon Australia Stationary Energy Plan: Australian Sustainable Energy. Melbourne Energy Institute, University of Melbourne.
6. D. Crawford, T. Jovanovic, M. O'Connor, A. Herr, J. Raison and T. Baynes (2012). AEMO 100% Renewable Energy Study: Potential for electricity generation in Australia from biomass in 2010, 2030 and 2050. Newcastle, Australia: CSIRO Energy Transformed Flagship.
7. T.V. Cutsem and C. Vournas (1998). Voltage Stability of Electric Power Systems. Boston: Kluwer Academic Publishers.
8. P. Kundur, N.J. Balu and M.G. Lauby (1994). Power System Stability and Control. New York: McGraw-Hill.
9. C.W. Taylor, N.J. Balu and D. Maratukulam (1994). Power System Voltage Stability. New York: McGraw-Hill.
10. P. Kundur, J. Paserba, V. Ajjarapu, G. Andersson, A. Bose, C. Canizares, N. Hatziargyriou, D. Hill, A. Stankovic, C. Taylor, T. Van Cutsem and V. Vittal (Aug. 2004). Definition and classification of power system stability. IEEE Transactions on Power Systems, 19(3): 1387-1401.
11. L.L. Lai, H.T. Zhang, C.S. Lai, F.Y. Xu and S. Mishra (2013). Investigation on July 2012 Indian Blackout, Proceedings of 2013 International Conference on Machine Learning and Cybernetics (ICMLC), 1(4): 92-97.
12. A.E.M. Operator (Dec, 2016). Black system South Australia, 28 Sep. 2016. 3rd Preliminary Report.
13. M. Pai (2012). Energy Function Analysis for Power System Stability. Springer Science & Business Media.
14. D. Gan, R.J. Thomas and R.D. Zimmerman (2000). Stability-constrained optimal power flow. IEEE Transactions on Power Systems, 15(2): 535-540.
15. D. Ruiz-Vega and M. Pavella (2003). A comprehensive approach to transient

stability control. I: Near optimal preventive control. IEEE Transactions on Power Systems, 18(4): 1446-1453.

16. A. Pizano-Martianez, C.R. Fuerte-Esquivel and D. Ruiz-Vega (2009). Global transient stability-constrained optimal power flow using an OMIB reference trajectory. IEEE Transactions on Power Systems, 25(1): 392-403.

17. R. Zárate-Miñano, T. Van Cutsem, F. Milano and A.J. Conejo (2009). Securing transient stability using time-domain simulations within an optimal power flow. IEEE Transactions on Power Systems, 25(1): 243-253.

18. A. Pizano-Martinez, C.R. Fuerte-Esquivel and D. Ruiz-Vega (2011). A new practical approach to transient stability-constrained optimal power flow. IEEE Transactions on Power Systems, 26(3): 1686-1696.

19. Y. Xue, T. Van Custem and M. Ribbens-Pavella (1989). Extended equal area criterion justifications, generalizations, applications. IEEE Transactions on Power Systems, 4(1): 44-52.

20. Y. Xu, Z.Y. Dong, K. Meng, J.H. Zhao and K.P. Wong (2012). A hybrid method for transient stability-constrained optimal power flow computation. IEEE Transactions on Power Systems, 27(4): 1769-1777.

21. Y. Xu, M. Yin, Z.Y. Dong, R. Zhang, D.J. Hill and Y. Zhang (2017). Robust dispatch of high wind power-penetrated power systems against transient instability. IEEE Transactions on Power Systems, 33(1): 174-186.

22. M. Pavella, D. Ernst and D. Ruiz-Vega (2012). Transient Stability of Power Systems: A Unified Approach to Assessment and Control. Springer Science & Business Media.

23. S. Gerbex, R. Cherkaoui and A.J. Germond (2001). Optimal location of multi-type FACTS devices in a power system by means of genetic algorithms. IEEE Transactions on Power Systems, 16(3): 537-544.

24. N. Yorino, E. El-Araby, H. Sasaki and S. Harada (2003). A new formulation for FACTS allocation for security enhancement against voltage collapse. IEEE Transactions on Power Systems, 18(1): 3-10.

25. Y.-C. Chang (2011). Multi-objective optimal SVC installation for power system loading margin improvement. IEEE Transactions on Power Systems, 27(2): 984-992.

26. E. Ghahremani and I. Kamwa (2012). Optimal placement of multiple-type FACTS devices to maximize power system loadability using a generic graphical user interface. IEEE Transactions on Power Systems, 28(2): 764-778.

27. M. Moghavvemi and O. Faruque (1998). Real-time contingency evaluation and ranking technique. IEE Proceedings – Generation, Transmission and Distribution, 145(5): 517-524.

28. C. Reis and F.M. Barbosa (2006). A comparison of voltage stability indices. Proceedings of 2006 IEEE Mediterranean Electrotechnical Conference (MELECON): 1007-1010.

29. M. Cupelli, C.D. Cardet and A. Monti (2012). Comparison of line voltage stability indices using dynamic real time simulation. Proceedings of 2012 3rd IEEE PES Innovative Smart Grid Technologies Europe (ISGT Europe): 1-8.

30. D. Shoup, J. Paserba and C. Taylor (2004). A survey of current practices for transient voltage dip/sag criteria related to power system stability. Proceedings of IEEE PES Power Systems Conference and Exposition: 1140-1147.

31. A. Tiwari and V. Ajjarapu (2011). Optimal allocation of dynamic VAR support using mixed integer dynamic optimization. IEEE Transactions on Power Systems, 26(1): 305-314.
32. Y. Xue, T. Xu, B. Liu and Y. Li (1999). Quantitative assessments for transient voltage security. Proceedings of 21st International Conference on Power Industry Computer Applications. Connecting Utilities. (PICA 99): 101-106.
33. Y. Dong, X. Xie, B. Zhou, W. Shi and Q. Jiang (2015). An Integrated High Side Var-Voltage Control Strategy to Improve Short-term Voltage Stability of Receiving-End Power Systems. IEEE Transactions on Power Systems, 31(3): 2105-2115.
34. Y. Xu, Z.Y. Dong, K. Meng, W.F. Yao, R. Zhang and K.P. Wong (2014). Multi-objective dynamic VAR planning against short-term voltage instability using a decomposition-based evolutionary algorithm. IEEE Transactions on Power Systems, 29(6): 2813-2822.
35. X. Taishan and X. Yusheng (2002). Quantitative assessments of transient frequency deviation acceptability [J]. Automation of Electric Power Systems, 26(19): 7-10.
36. A.G. Phadke and J.S. Thorp (2008). Synchronized Phasor Measurements and Their Applications. Springer.
37. C. Wan, Z. Xu, P. Pinson, Z.Y. Dong and K.P. Wong (May 2014). Probabilistic forecasting of wind power generation using extreme learning machine. IEEE Transactions on Power Systems, 29(3): 1033-1044.
38. W.C. Kong, Z.Y. Dong, Y.W. Jia, D.J. Hill, Y. Xu and Y. Zhang (Jan. 2019). Short-term residential load forecasting based on LSTM recurrent neural network. IEEE Transactions on Smart Grid, 10(1): 841-851.
39. C. Ramos and C.C. Liu (Mar.-Apr, 2011). AI in power systems and energy markets introduction. IEEE Intelligent Systems, 26(2): 5-8.
40. Z. Vale, T. Pinto, I. Praca and H. Morais (Mar.-Apr, 2011). MASCEM: Electricity markets simulation with strategic agents. IEEE Intelligent Systems, 26(2): 9-17.
41. Y.F. Guan and M. Kezunovic (Mar.-Apr. 2011). Grid monitoring and market risk management. IEEE Intelligent Systems, 26(2): 18-27.
42. J. Ferreira, S. Ramos, Z. Vale and J. Soares (Mar.-Apr. 2011). A data-mining-based methodology for transmission expansion planning. IEEE Intelligent Systems, 26(2): 28-37.
43. C.F.M. Almeida and N. Kagan (Mar.-Apr. 2011). Using genetic algorithms and fuzzy programming to monitor voltage sags and swells. IEEE Intelligent Systems, 26(2): 46-53.
44. Z.Y. Dong, Y. Xu, P. Zhang and K.P. Wong (Jul.-Aug, 2013). Using IS to assess an electric power systems real-time stability. IEEE Intelligent Systems, 28(4): 60-66.
45. L. Wehenkel, V. Cutsem and M. Ribbens-Pavella (1989). An artificial intelligence framework for on-line transient stability assessment of power systems. IEEE Transactions on Power Systems, 4(2): 789-800.
46. L. Wehenkel, M. Pavella, E. Euxibie and B. Heilbronn (1994). Decision tree based transient stability method a case study. IEEE Transactions on Power Systems, 9(1): 459-469.

47. C.A. Jensen, M.A. El-Sharkawi and R.J. Marks (2001). Power system security assessment using neural networks: Feature selection using Fisher discrimination. IEEE Transactions on Power Systems, 16(4): 757-763.
48. E.M. Voumvoulakis, A.E. Gavoyiannis and N.D. Hatziargyriou (2006). Decision trees for dynamic security assessment and load shedding scheme. 2006 Power Engineering Society General Meeting, 1-9: 3303-3309.
49. K. Sun, S. Likhate, V. Vittal, V.S. Kolluri and S. Mandal (Nov. 2007). An On-line dynamic security assessment scheme using phasor measurements and decision trees. IEEE Transactions on Power Systems, 22(4): 1935-1943.
50. Y. Xu, Z.Y. Dong, J.H. Zhao, K. Meng and K.P. Wong (2010). Transient stability assessment based on data-structure analysis of operating point space. IEEE Power and Energy Society General Meeting, 2010.
51. J.H. Zhao, Z.Y. Dong and P. Zhang (2007). Mining complex power networks for blackout prevention. KDD-2007 Proceedings of the Thirteenth ACM SIGKDD International Conference on Knowledge Discovery and Data Mining. pp. 986-994.
52. I. Kamwa, R. Grondin and L. Loud (Aug. 2001). Time-varying contingency screening for dynamic security assessment using intelligent-systems techniques. IEEE Transactions on Power Systems, 16(3): 526-536.
53. I. Kamwa, S.R. Samantaray and G. Joos (Sep. 2010). Catastrophe predictors from ensemble decision-tree learning of wide-area severity indices. IEEE Transactions on Smart Grid, 1(2): 144-158.
54. I. Kamwa, S.R. Samantaray and G. Joos (Feb. 2009). Development of rule-based classifiers for rapid stability assessment of wide-area post-disturbance records. IEEE Transactions on Power Systems, 24(1): 258-270.
55. N. Amjady and S.F. Majedi (2007). Transient stability prediction by a hybrid intelligent system. IEEE Transactions on Power Systems, 22(3): 1275-1283.
56. N. Amjady and S. Banihashemi (2010). Transient stability prediction of power systems by a new synchronism status index and hybrid classifier. IET Generation, Transmission & Distribution, 4(4): 509-518.
57. A.D. Rajapakse, F. Gomez, K. Nanayakkara, P.A. Crossley and V.V. Terzija (2009). Rotor angle instability predction using post-disturbance voltage trajectories. IEEE Transactions on Power Systems, 25(2): 947-956.
58. V. Terzija, G. Valverde, D.Y. Cai, P. Regulski, V. Madani, J. Fitch, S. Skok, M.M. Begovic and A. Phadke (Jan. 2011). Wide-area monitoring, protection, and control of future electric power networks. Proceedings of the IEEE, 99(1): 80-93.
59. K. Kirihara, K.E. Reinhard, A.K. Yoon and P.W. Sauer (2014). Investigating synchrophasor data quality issues. Proceedings of 2014 Power and Energy Conference at Illinois (PECI): 1-4.
60. F. Aminifar, M. Fotuhi-Firuzabad, M. Shahidehpour and A. Safdarian (Sep. 2012). Impact of WAMS malfunction on power system reliability assessment. IEEE Transactions on Smart Grid, 3(3): 1302-1309.
61. Y. Wang, W.Y. Li and J.P. Lu (Jul. 2010). Reliability analysis of wide-area measurement system. IEEE Transactions on Power Delivery, 25(3): 1483-1491.
62. K.E. Martin (2015). Synchrophasor measurements under the IEEE standard C37. 118.1-2011 with amendment C37. 118.1 a. IEEE Transactions on Power Delivery, 30(3): 1514-1522.

63. H. Farhangi (Jan.-Feb. 2010). The path of the smart grid. IEEE Power & Energy Magazine, 8(1): 18-28.

64. E. SmartGrids (2006). Vision and Strategy for Europe's Electricity Networks of the Future. European Commission, 2006.

65. Z. Dong, K.P. Wong, K. Meng, F.J. Luo, F. Yao and J.H. Zhao (2010). Wind power impact on system operations and planning. IEEE Power and Energy Society General Meeting 2010.

66. H. Banakar, C. Luo and B.T. Ooi (Feb. 2008). Impacts of wind power minute-to-minute variations on power system operation. IEEE Transactions on Power Systems, 23(1): 150-160.

67. T.Y. Guo and J.V. Milanovic (2013). The Effect of quality and availability of measurement signals on accuracy of on-line prediction of transient stability using decision tree method. 2013 4th IEEE/Pes Innovative Smart Grid Technologies Europe (Isgt Europe).

68. M. He, V. Vittal and J.S. Zhang (May 2013). On-line dynamic security assessment with missing PMU measurements: A data mining approach. IEEE Transactions on Power Systems, 28(2): 1969-1977.

69. H. Sawhney and B. Jeyasurya (Aug. 2006). A feed-forward artificial neural network with enhanced feature selection for power system transient stability assessment. Electric Power Systems Research, 76(12): 1047-1054.

70. Y. Xu, Z.Y. Dong, L. Guan, R. Zhang, K.P. Wong and F.J. Luo (Aug. 2012). Preventive dynamic security control of power systems based on pattern discovery technique. IEEE Transactions on Power Systems, 27(3): 1236-1244.

71. L.S. Moulin, A.P.A. da Silva, M.A. El-Sharkawi and R.J. Marks (May 2004). Support vector machines for transient stability analysis of large-scale power systems. IEEE Transactions on Power Systems, 19(2): 818-825.

72. Y. Xu, Z.Y. Dong, J.H. Zhao, P. Zhang and K.P. Wong (Aug. 2012). A reliable intelligent system for real-time dynamic security assessment of power systems. IEEE Transactions on Power Systems, 27(3): 1253-1263.

73. R.S. Diao, K. Sun, V. Vittal, R.J. O'Keefe, M.R. Richardson, N. Bhatt, D. Stradford and S.K. Sarawgi (May 2009). Decision tree-based on-line voltage security assessment using PMU measurements. IEEE Transactions on Power Systems, 24(2): 832-839.

74. C.X. Liu, K. Sun, Z.H. Rather, Z. Chen, C.L. Bak, P. Thogersen and P. Lund (Mar. 2014). A systematic approach for dynamic security assessment and the corresponding preventive control scheme based on decision trees. IEEE Transactions on Power Systems, 29(2): 717-730.

75. M. He, J.S. Zhang and V. Vittal (Nov. 2013). Robust on-line dynamic security assessment using adaptive ensemble decision-tree learning. IEEE Transactions on Power Systems, 28(4): 4089-4098.

76. K. Morison (2006). On-line dynamic security assessment using intelligent systems. 2006 Power Engineering Society General Meeting, 1-9: 3292-3296.

77. Y. Xu, Z. Dong, K. Meng, R. Zhang and K. Wong (2011). Real-time transient stability assessment model using extreme learning machine. IET Generation, Transmission & Distribution, 5(3): 314-322.

78. U. Manual (2009). OptiLoad (v1. 0b). The Hong Kong Polytechnic University, Kowloon, Hong Kong.

79. J. Zhu (2015). Optimization of Power System Operation, John Wiley & Sons.

80. T. wen Wang, L. Guan and Y. Zhang (2008). A modified pattern recognition algorithm and its application in power system transient stability assessment. Proceedings on 2008 IEEE PES General Meeting: 1-7.
81. K. Kira and L.A. Rendell (1992). A practical approach to feature selection. Machine Learning Proceedings 1992, Elsevier. pp. 249-256.
82. M. Robnik-Šikonja and I. Kononenko (2003). Theoretical and empirical analysis of ReliefF and Relief. Machine Learning, 53(1-2): 23-69.
83. G.B. Huang, Q.Y. Zhu and C.K. Siew (Dec. 2006). Extreme learning machine: Theory and applications. Neurocomputing, 70(1-3): 489-501.
84. N.Y. Liang, G.B. Huang, P. Saratchandran and N. Sundararajan (Nov. 2006). A fast and accurate on-line sequential learning algorithm for feedforward networks. IEEE Transactions on Neural Networks, 17(6): 1411-1423.
85. G.B. Huang, L. Chen and C.K. Siew (Jul. 2006). Universal approximation using incremental constructive feedforward networks with random hidden nodes. IEEE Transactions on Neural Networks, 17(4): 879-892.
86. G.B. Huang, D.H. Wang and Y. Lan (Jun. 2011). Extreme learning machines: A survey. International Journal of Machine Learning and Cybernetics, 2(2): 107-122.
87. R.M. Golden (Sep. 1997). Neural networks: A comprehensive foundation – Haykin, S. Journal of Mathematical Psychology, 41(3): 287-292.
88. S.A. Corne (1996). Artificial neural networks for pattern recognition. Concepts in Magnetic Resonance, 8(5): 303-324.
89. L. Breiman (1984). Classification and Regression Trees. Belmont, Calif. Wadsworth International Group.
90. J. Han, J. Pei and M. Kamber (2011). Data Mining: Concepts and Techniques, Elsevier.
91. J. Han, M. Kamber and J. Pei (2012). Data Mining: Concepts and Techniques, 3rd edition, pp. 1-703.
92. F.R. Gomez, A.D. Rajapakse, U.D. Annakkage and I.T. Fernando (2011). Support vector machine-based algorithm for post-fault transient stability status prediction using synchronized measurements. IEEE Transactions on Power Systems, 26(3): 1474-1483.
93. Y.-H. Pao, G.-H. Park and D.J. Sobajic (1994). Learning and generalization characteristics of the random vector functional-link net. Neurocomputing, 6(2): 163-180.
94. Y. Jia, K. Meng and Z. Xu (2014). Nk-induced cascading contingency screening. IEEE Transactions on Power Systems, 30(5): 2824-2825.
95. S.Chai, Z. Xu and W.K. Wong (2015). Optimal granule-based PIs construction for solar irradiance forecast. IEEE Transactions on Power Systems, 31(4): 3332-3333.
96. Y. Ren, P.N. Suganthan, N. Srikanth and G. Amaratunga (2016). Random vector functional link network for short-term electricity load demand forecasting. Information Sciences, 367: 1078-1093.
97. L. Zhang and P.N. Suganthan (2016). A comprehensive evaluation of random vector functional link networks. Information Sciences, 367: 1094-1105.
98. L. Zhang and P.N. Suganthan (2016). A survey of randomized algorithms for training neural networks. Information Sciences, 364: 146-155.

99. W.F. Schmidt, M.A. Kraaijveld and R.P. Duin (1992). Feed Forward Neural Networks with Random Weights. Proceedings of International Conference on Pattern Recognition: 1-1.

100. K. Meng, Z.Y. Dong, K.P. Wong, Y. Xu and F.J. Luo (May 2010). Speed-up the computing efficiency of power system simulator for engineering-based power system transient stability simulations. Iet Generation Transmission & Distribution, 4(5): 652-661.

101. T. Jain, L. Srivastava and S.N. Singh (Nov. 2003). Fast voltage contingency screening using radial basis function neural network. IEEE Transactions on Power Systems, 18(4): 1359-1366.

102. L.K. Hansen and P. Salamon (Oct. 1990). Neural network ensembles. IEEE Transactions on Pattern Analysis and Machine Intelligence, 12(10): 993-1001.

103. L. Breiman (2001). Random forests. Machine Learning, 45(1): 5-32.

104. Y. Ren, L. Zhang and P.N. Suganthan (2016). Ensemble classification and regression-recent developments, applications and future directions. IEEE Computational Intelligence Magazine, 11(1): 41-53.

105. V. Vittal, D. Martin, R. Chu, J. Fish, J.C. Giri, C.K. Tang, F.E. Villaseca and R. Farmer (1992). Transient stability test systems for direct stability methods. IEEE Transactions on Power Systems, 7(1): 37.

106. Y. Mansour, E. Vaahedi and M.A. El-Sharkawi (1997). Dynamic security contingency screening and ranking using neural networks. IEEE Transactions on Neural Networks, 8(4): 942-950.

107. Z.H. Zhou, J.X. Wu and W. Tang (May 2002). Ensembling neural networks: Many could be better than all. Artificial Intelligence, 137(1-2): 239-263.

108. K. Miettinen (2012). Non-linear Multiobjective Optimization, Springer Science & Business Media.

109. J. Han and M. Kamber (2001). Data mining: Concepts and Techniques. San Francisco: Morgan Kaufmann Publishers.

110. S. Agrawal, B. Panigrahi and M.K. Tiwari (2008). Multiobjective particle swarm algorithm with fuzzy clustering for electrical power dispatch. IEEE Transactions on Evolutionary Computation, 12(5): 529-541.

111. B. Zhou, K. Chan, T. Yu and C. Chung (2013). Equilibrium-inspired multiple group search optimization with synergistic learning for multiobjective electric power dispatch. IEEE Transactions on Power Systems, 28(4): 3534-3545.

112. M. Seyedi (2009). Evaluation of the DFIG wind turbine built-in model in PSS/E. Master's thesis. Gothenburg, Sweden: Chalmers University of Technology.

113. A. Estabrooks, T. Jo and N. Japkowicz (2004). A multiple resampling method for learning from imbalanced data sets. Computational Intelligence, 20(1): 18-36.

114. E.A. Committee (2008). Smart grid: Enabler of the new energy economy. A Report by the Electricity Advisory Committee.

115. G.R. Hancock, M.S. Butler and M.G. Fischman (1995). On the problem of two-dimensional error scores: Measures and analyses of accuracy, bias, and consistency. Journal of Motor Behavior, 27(3): 241-250.

116. J.D. De Leon and C.W. Taylor (2002). Understanding and solving short-term voltage stability problems. Proceedings of IEEE Power Engineering Society Summer Meeting, 2: 745-752.

117. R. Zhang, Y. Xu, Z.Y. Dong, K. Meng and Z. Xu (2011). Intelligent systems for power system dynamic security assessment: Review and classification. Proceedings of 2011 4th International Conference on Electric Utility Deregulation and Restructuring and Power Technologies (DRPT): 134-139.

118. E. Karapidakis and N. Hatziargyriou (2002). On-line preventive dynamic security of isolated power systems using decision trees. IEEE Transactions on Power Systems, 17(2): 297-304.

119. K. Mei and S.M. Rovnyak (2004). Response-based decision trees to trigger one-shot stabilizing control. IEEE Transactions on Power Systems, 19(1): 531-537.

120. E.M. Voumvoulakis and N.D. Hatziargyriou (2008). Decision trees-aided self-organized maps for corrective dynamic security. IEEE Transactions on Power Systems, 23(2): 622-630.

121. K. Niazi, C. Arora and S. Surana (2003). A hybrid approach for security evaluation and preventive control of power systems. Proceedings of 2003 National Power Engineering Conference (PECON): 193-199.

122. I. Genc, R. Diao, V. Vittal, S. Kolluri and S. Mandal (2010). Decision tree-based preventive and corrective control applications for dynamic security enhancement in power systems. IEEE Transactions on Power Systems, 25(3): 1611-1619.

123. A.K.C. Wong and Y. Wang (2003). Pattern discovery: A data driven approach to decision support. IEEE Transactions on Systems, Man, and Cybernetics, Part C (Applications and Reviews), 33(1): 114-124

124. T. Chau and A.K. Wong (1999). Pattern discovery by residual analysis and recursive partitioning. IEEE Transactions on Knowledge and Data Engineering, 11(6): 833-852.

125. T.B. Nguyen and M. Pai (2003). Dynamic security-constrained rescheduling of power systems using trajectory sensitivities. IEEE Transactions on Power Systems, 18(2): 848-854.

126. R.D. Zimmerman, C.E. Murillo-Sánchez and R.J. Thomas (2010). MATPOWER: Steady-state operations, planning, and analysis tools for power systems research and education. IEEE Transactions on Power Systems, 26(1): 12-19.

127. D. Steinberg and M. Golovnya (2006). CART 6.0 User's Manual. Salford Systems, San Diego, CA.

128. Y. Xu, L. Guan, Z.Y. Dong and K.P. Wong (2010). Transient stability assessment on China Southern Power Grid System with an improved pattern discovery-based method. Proceedings of 2010 International Conference on Power System Technology: 1-6.

129. C.M. Goldie (1995). Theoretical Probability for Applications. JSTOR.

130. S. Rovnyak, S. Kretsinger, J. Thorp and D. Brown (Aug. 1994). Decision trees for real-time transient stability prediction. IEEE Transactions on Power Systems, 9(3): 1417-1426.

131. E.M. Voumvoulakis and N.D. Hatziargyriou (May 2010). A particle swarm optimization method for power system dynamic security control. IEEE Transactions on Power Systems, 25(2): 1032-1041.

132. S.M. Halpin, K.A. Harley, R.A. Jones and L.Y. Taylor (2008). Slope-permissive under-voltage load shed relay for delayed voltage recovery mitigation. IEEE Transactions on Power Systems, 23(3): 1211-1216.

133. H. Bai and V. Ajjarapu (2010). A novel on-line load shedding strategy for mitigating fault-induced delayed voltage recovery. IEEE Transactions on Power Systems, 26(1): 294-304.

134. S. Dasgupta, M. Paramasivam, U. Vaidya and V. Ajjarapu (2014). Entropy-based metric for characterization of delayed voltage recovery. IEEE Transactions on Power Systems, 30(5): 2460-2468.

135. Y. Dong, X. Xie, K. Wang, B. Zhou and Q. Jiang (2017). An emergency-demand-response based under speed load shedding scheme to improve short-term voltage stability. IEEE Transactions on Power Systems, 32(5): 3726-3735.

136. M. Begovic, D. Novosel, D. Karlsson, C. Henville and G. Michel (2005). Wide-area protection and emergency control. Proceedings of the IEEE, 93(5): 876-891.

137. R. Zhang, Y. Xu, Z.Y. Dong and D.J. Hill (2012). Feature selection for intelligent stability assessment of power systems. Proceedings of 2012 IEEE PES General Meeting: 1-7.

138. J.C. Cepeda, J.L. Rueda, D.G. Colomé and I. Erlich (2015). Data-mining-based approach for predicting the power system post-contingency dynamic vulnerability status. International Transactions on Electrical Energy Systems, 25(10): 2515-2546.

139. J.C. Cepeda, J.L. Rueda, D.G. Colomé and D.E. Echeverría (2014). Real-time transient stability assessment based on centre-of-inertia estimation from phasor measurement unit records. IET Generation, Transmission & Distribution, 8(8): 1363-1376.

140. J.C. Cepeda, J.L. Rueda, I. Erlich and D.G. Colomé (2012). Probabilistic approach-based PMU placement for real-time power system vulnerability assessment. Proceedings of 2012 IEEE PES Innovative Smart Grid Technologies Europe (ISGT Europe): 1-8.

141. M. Glavic, D. Novosel, E. Heredia, D. Kosterev, A. Salazar, F. Habibi-Ashrafi and M. Donnelly (2012). See it fast to keep calm: Real-time voltage control under stressed conditions. IEEE Power and Energy Magazine, 10(4): 43-55.

142. E.G. Potamianakis and C.D. Vournas (2006). Short-term voltage instability: Effects on synchronous and induction machines. IEEE Transactions on Power Systems, 21(2): 791-798.

143. A.S. Meliopoulos, G. Cokkinides and G. Stefopoulos (2006). Voltage stability and voltage recovery: Load dynamics and dynamic VAR sources. Proceedings of 2006 IEEE PES General Meeting. p. 8.

144. J. Liu, Y. Xu, Z.Y. Dong and K.P. Wong (2017). Retirement-driven dynamic VAR planning for voltage stability enhancement of power systems with high-level wind power. IEEE Transactions on Power Systems, 33(2): 2282-2291.

145. S. Dasgupta, M. Paramasivam, U. Vaidya and V. Ajjarapu (2013). Real-time monitoring of short-term voltage stability using PMU data. IEEE Transactions on Power Systems, 28(4): 3702-3711.

146. L. Zhu, C. Lu and Y. Sun (2015). Time Series Shapelet Classification Based On-line Short-term Voltage Stability Assessment. IEEE Transactions on Power Systems, 31(2): 1430-1439.

147. L. Zhu, C. Lu, Z.Y. Dong and C. Hong (2017). Imbalance learning machine-based power system short-term voltage stability assessment. IEEE Transactions on Industrial Informatics, 13(5): 2533-2543.
148. Y. Xu, R. Zhang, J. Zhao, Z. Dong, D. Wang, H. Yang and K. Wong (2015). Assessing short-term voltage stability of electric power systems by a hierarchical intelligent system. IEEE Transactions on Neural Networks and Learning Systems, 27(8): 1686-1696.
149. K. Kira and L.A. Rendell (1992). A practical approach to feature-selection. Machine Learning. Proceedings of International Conference on Machine Learning (ICML'92): 249-256.
150. G.B. Huang, H.M. Zhou, X.J. Ding and R. Zhang (Apr. 2012). Extreme learning machine for regression and multiclass classification. IEEE Transactions on Systems Man and Cybernetics Part B – Cybernetics, 42(2): 513-529.
151. P. Siemens (2011). PSS/E 33.0 Program Application Guide: Volume II, Siemens PTI: Schenectady, NY, USA.
152. R. Zhang, Y. Xu, Z.Y. Dong and K.P. Wong (2016). Measurement-based dynamic load modeling using time-domain simulation and parallel-evolutionary search. IET Generation, Transmission & Distribution, 10(15): 3893-3900.
153. Z. Guo, W. Zhao, H. Lu and J. Wang (2012). Multi-step forecasting for wind speed using a modified EMD-based artificial neural network model. Renewable Energy, 37(1): 241-249.
154. K. Deb, A. Pratap, S. Agarwal and T. Meyarivan (2002). A fast and elitist multiobjective genetic algorithm: NSGA-II. IEEE Transactions on Evolutionary Computation, 6(2): 182-197.
155. I.S. Association (2011). IEEE standard for synchrophasor data transfer for power systems. IEEE Std C, 37: 18-20.
156. A. Khamis, Y. Xu, Z.Y. Dong and R. Zhang (2016). Faster detection of microgrid islanding events using an adaptive ensemble classifier. IEEE Transactions on Smart Grid, 9(3): 1889-1899.
157. I. Kamwa, S. Samantaray and G. Joós (2011). On the accuracy versus transparency tradeoff of data-mining models for fast-response PMU-based catastrophe predictors. IEEE Transactions on Smart Grid, 3(1): 152-161.
158. Y.-Y. Hong and S.-F. Wei (2010). Multiobjective underfrequency load shedding in an autonomous system using hierarchical genetic algorithms. IEEE Transactions on Power Delivery, 25(3): 1355-1362.
159. L. Sigrist, I. Egido, E.F. Sanchez-Ubeda and L. Rouco (2009). Representative operating and contingency scenarios for the design of UFLS schemes. IEEE Transactions on Power Systems, 25(2): 906-913.
160. X. Lin, H. Weng, Q. Zou and P. Liu (2008). The frequency closed-loop control strategy of islanded power systems. IEEE Transactions on Power Systems, 23(2): 796-803.
161. C.-T. Hsu, M.-S. Kang and C.-S. Chen (2005). Design of adaptive load shedding by artificial neural networks. IEE Proceedings – Generation, Transmission and Distribution, 152(3): 415-421.
162. W. Da, X. Yusheng and X. Taishan (2009). Optimization and coordination of fault-driven load shedding and trajectory-driven load shedding. Automation of Electric Power Systems, 2009(13): 13.

163. Y. Xue (2007). Towards space-time cooperative defence framework against blackouts in China. Proceedings of 2007 IEEE PES General Meeting: 1-6.

164. Y. Xue (1998). Fast analysis of stability using EEAC and simulation technologies. Proceedings of 1998 International Conference on Power System Technology (PowerCon 1998), 1: 12-16.

165. Y.J. Fang and Y. Xue (2000). An on-line pre-decision based transient stability control system for the Ertan power system. Proceedings of 2000 International Conference on Power System Technology (PowerCon 2000), 1: 287-292.

166. X. Yusheng, W. Da and W. Fushuan (2009). A review on optimization and coordination of emergency control and correction control. Automation of Electric Power Systems, 33(12): 1-7.

167. Z. Xu, J. Ostergaard and M. Togeby (2010). Demand as frequency controlled reserve. IEEE Transactions on Power Systems, 26(3): 1062-1071.

168. Y. Wang, I.R. Pordanjani and W. Xu (2011). An event-driven demand response scheme for power system security enhancement. IEEE Transactions on Smart Grid, 2(1): 23-29.

169. R.-F. Chang, C.-N. Lu and T.-Y. Hsiao (2005). Prediction of frequency response after generator outage using regression tree. IEEE Transactions on Power Systems, 20(4): 2146-2147.

170. Y.-H. Chen and Y.-S. Xue (2000). Optimal algorithm for regional emergency control. Electric Power, 33(1): 44-48.

171. A.N. Baraldi and C.K. Enders (Feb. 2010). An introduction to modern missing data analyses. Journal of School Psychology, 48(1): 5-37.

172. M. Saar-Tsechansky and F. Provost (Jul. 2007). Handling missing values when applying classification models. Journal of Machine Learning Research, 8: 1625-1657.

173. Q. Li, Y. Xu, C. Ren and J. Zhao (2019). A hybrid data-driven method for on-line power system dynamic security assessment with incomplete PMU measurements. Proceedings of 2019 IEEE PES General Meeting: 1-6.

174. B. Gou (2008). Generalized integer linear programming formulation for optimal PMU placement. IEEE transactions on Power Systems, 23(3): 1099-1104.

175. S. Chakrabarti and E. Kyriakides (2008). Optimal placement of phasor measurement units for power system observability. IEEE Transactions on Power Systems, 23(3): 1433-1440.

176. S. Chakrabarti, E. Kyriakides and D.G. Eliades (2008). Placement of synchronized measurements for power system observability. IEEE Transactions on Power Delivery, 24(1): 12-19.

177. F. Aminifar, A. Khodaei, M. Fotuhi-Firuzabad and M. Shahidehpour (2009). Contingency-constrained PMU placement in power networks. IEEE Transactions on Power Systems, 25(1): 516-523.

178. N. Amjady and S.A. Banihashemi (Apr. 2010). Transient stability prediction of power systems by a new synchronism status index and hybrid classifier. IET Generation Transmission & Distribution, 4(4): 509-518.

179. D.P. Tools (2013). Transient Security Assessment Tool User Manual, Powertech Labs, Inc., Surrey, British Columbia.

180. I. Goodfellow, J. Pouget-Abadie, M. Mirza, B. Xu, D. Warde-Farley, S. Ozair, A. Courville and Y. Bengio (2014). Generative Adversarial Nets. Advances in Neural Information Processing Systems: 2672-2680.

181. D. Pathak, P. Krahenbuhl, J. Donahue, T. Darrell and A.A. Efros (2016). Context encoders: Feature learning by inpainting. Proceedings of IEEE Conference on Computer Vision and Pattern Recognition: 2536-2544.
182. Y. Chen, Y. Wang, D. Kirschen and B. Zhang (2018). Model-free renewable scenario generation using generative adversarial networks. IEEE Transactions on Power Systems, 33(3): 3265-3275.
183. D.P. Kingma and J. Ba (2014). Adam: A method for stochastic optimization. arXiv Preprint arXiv:1412.6980.
184. J.S. Armstrong and F. Collopy (1992). Error measures for generalizing about forecasting methods: Empirical comparisons. International Journal of Forecasting, 8(1): 69-80.
185. G.E. Batista and M.C. Monard (2003). An analysis of four missing data treatment methods for supervised learning. Applied Artificial Intelligence, 17(5-6): 519-533.

Index